GOD EMBRACES

SCRIPTURE, SCIENCE & MIRACLES

My Personal Journey
From Skepticism to Belief

Dr. H. Ed Logue, M.D.

ISBN: 978-0-9990047-1-5

DEDICATION

This book is dedicated to God, and to His believers who have assisted me in ways to bring the book to completion.

A LIFE OF MIRACLES

This is an uplifting and heartwarming compilation of miracles that I or others have experienced firsthand, and desire to share with you, that reveal the power, love and grace of our living God.

I invite you to travel along with me on my journey from science based skepticism to an undeniable belief that God is a loving, intelligent, creative being, who is involved in our day to day lives, as demonstrated to me (and others) through the events and occurrences shared here.

CONTENTS

OPENING PRAYER

Almighty God, creator and sustainer, please accept this prayer, to give You thanks and praises for the privilege and blessing of this life, and as a plea from me, a former "doubting Thomas," acutely aware of my shortcomings, but being a human repository of significant miracles that should not fade with the flesh but be liberated to any who read or hear them. Hopefully, many will accept and understand the odds alone are mathematical proof of a creator, as we know You to be, and undeniably beyond the limits of human calculability, not even close to emulating. Therefore, assist my earnest hand with Your divine touch so as to reach those in need of proof to become aware of the living active intervening God You have shown Yourself to be. May my readers live inch by inch closer to You and find the bountiful joy of reaching your presence.

There is no peace or satisfaction of purpose stronger than this truth.

FIRST HISTORICAL MIRACLES

1. God discovers and develops Himself.
2. God creates the universe.
3. God creates life.
4. God designs and creates the human species.

Relevant to humans these four must be the historical first and most awesome miracles, unparalleled and unmatched by anything else known to man or woman. Perhaps human birth and procreation is number 5.

MIRACLES FORETASTE

Introduction Part 1:

Miracles are more frequent and personally directed than generally recognized or acknowledged. Often, they are powerful in message or action. The odds of them being coincidental are patently impossible. The following is one such example. It is a foretaste of several such miracles, and a host of others, as well as a deep look at the science of religion, creation and miracles.

This amazing miracle I experienced while in meditation (shared in the following chapter) should be read with appropriate awe for its drama, undoubtedly written and flawlessly orchestrated from the beyond.

You are invited into my life of miracles. A full analysis and my interpretive message of this initial miracle follows shortly in an upcoming chapter.

Take the foregoing to heart. Pray to God that each of your five senses will be open to know a miracle when it happens. Notice its intimacy for you, or others, and take solace and comfort from it.

I asked God to show me His reality. He did! The first miracle you are about to read was an early and compelling answer followed by an array of others.

You will seek Me and find Me when you seek Me with all your heart.[1]

[1] Jeremiah 29:13

MIRACLE TYPES

The first miracle you are about to read is orchestrated and fine-tuned.

Some are by dreams.

Some are by healing.

Some are by instant life saving knowledge.

Some are by a feeling of discomfort, given to us by the Holy Spirit.

Some are by a simple meaningful sign.

Some are in response to prayer, individual or collective.

Some are God's pleasure to give, yet completely unrecognizable by the recipient.

Large or small, He does them all.

A MULTI-PURPOSE MIRACLE

In the 1960's I was in the practice of general medicine. One of my patients mentioned to me, since I was familiar with hypnosis and things of the mind, I might be interested in coming to one of their spiritual meetings. I asked the nature of the meeting. She described the meeting as a nondenominational gathering of like-minded Christians meeting to meditate on spirituality. I agreed to attend, as a courtesy to a patient.

I arrived on time at a modest residential apartment complex. It was without fancy trimmings or a guarded gate. A pleasant lady, the hostess, appearing to be approximately sixty-five years old invited me in. A couple of people were already there and a few more arrived shortly. In all, I, the host and eight others were present. I sat to the right, next to the door. To my right one person sat and the host in the corner. On the right wall two persons sat and one in the corner with four others to her right on the window wall which faced me. It was past sundown and conversation was sparse. I do not recall if introductions were made. Nothing significant was mentioned regarding anyone present. Someone took the lead and mentioned Scripture would be read. Fifteen minutes of meditation would follow after which we should share the thoughts or experiences we had during meditation. This seemed strange and caused discomfort to me. The following Scripture was read, "Ask, and it shall be given you; seek, and ye shall find; knock and it shall be opened unto you." We were then asked to meditate.

My meditation consisted of sitting with my eyes closed most of the time, but occasionally I would open my eyes and look around the room. I really felt quite out of place, but I did as we were instructed except for the opening of my eyes. I wondered what was supposed to be happening. I would turn my head from right to left, sometimes with my eyes closed and sometimes open.

A surprise came to me. I suddenly saw a rustic wooden pavilion as I looked at the lady in the corner to my right on the window wall. As I looked at the last person to my left, I saw pain, discomfort and sorrow. It was as plain as if I were seeing a physical object. Otherwise, she was no different than any other time I had looked at her. I thought little about it. On another sweep I saw a man standing just behind the host, while in reality there was no man there. After that, each time I looked, these visions became more clear and vivid. I was amazed, even bewildered, that I was having visions as real as the people in the room.

Eventually the leader announced the end of the meditation and asked for volunteers to give comments about what transpired during meditation. I sat unrevealing while two or three others made comments. The only one I remember is that the lady in the middle facing me mentioned that when she looked at me, she saw a man behind me, distinguished looking with a full beard. She declared to me that he was my guardian angel and was always with me. I expressed some satisfaction and thankfulness. When several had spoken, I decided perhaps I should at least mention what I had seen. It would be unfair to accept an invitation for a declared purpose and not acknowledge my experience.

First, I mentioned to the lady in the middle in the corner that each time I looked at her I saw a rustic camp pavilion, as it might be on a riverbank. With exuberant joy, she exclaimed gleefully, "That is where my husband proposed to me." She said they visit there often as it belongs to her husband's family. She was absolutely animated and delighted. It was "their favorite place." She wondered openly, but rhetorically, how did I know? She, as I, knew we had no answer.

The feeling I had at that point was surprise and wonderment. How could this possibly be? I knew nothing of her previously and certainly had no reason to think of a wilderness or riverside camp. It had however made her feel joy.

I then turned to the lady to my far left and said, I'm sorry but every time I looked at you, I saw sorrow, pain and discord. It appeared that this was marital. She started crying and said, "Yes, but no one can know. That must be kept secret because my husband is a minister and he will lose his job if that gets out."

By then I was feeling overwhelmed. Something was happening that I could not explain. This was a very new experience for me. The odds that I would have picked up on the fish camp seemed to me to be at least 1000 to 1. Likewise, with the second lady, 1000 to 1. The odds that this would happen twice in one evening to someone with no experience at all in this kind of situation, you would multiply the 1000 times 1000. The odds would be a million to one that could happen by chance, or so it seemed to me.

I then turned to the lady on my right, the hostess, and said, "Ma'am, every time I looked at you, I saw a man standing behind you. He stood to your right. He appeared to be 40 to 45 years of age and he was in a black suit and tie. He was talking to me without moving his lips. I could hear him plainly. I could see him as clearly as I see you. He appeared to be about 5'8" to 5'10" tall.

The lady spoke, "That could be my husband. He died at 43 years old, 30 years ago. And he was 5'10"."

Until that moment I did not know that she was widowed. I perceived the need to proceed with the rest of what I had seen. "I watched as he pulled a pocket watch from his suit, looked me in the eye and with a deliberate focus, took his forefinger, opened the lid of the watch and pointed to the inside of the cover emphasizing the presence of a message. He wanted me to acknowledge that I would mention that message."

She commented that did not remember him having a pocket watch. This did not faze me as I was already bewildered. I responded, "I am reporting what I saw and heard."

Also, I saw a scar, though not its location. I was as sure of it as I was of the other lady's sorrow. It had to be there.

The hostess however, said, "I'm sorry, he did not have a scar."

I mentioned that he also had a beard, but again her response was, "I'm sorry, he did not wear a beard."

I thought, well now, this is confusing, first a build-up then a let-down. Perhaps this shows me none of this can be truly meaningful. Then came the moment to finish the encounter. I told her the last thing he said before he disappeared, "There was one more thing that he had plainly said to me without moving his lips. He said, 'Tell her that the one thing she continues to worry about never bothered or concerned me. She should let it go, forget it, and be at peace.'"

At this point, she instantly burst into tears. Through her sobbing she said "I have longed for this and prayed for this ever since he died." As she cried, she suddenly stopped crying and looked up and said, "Oh my, he did have a watch and I do remember the inscription and it is meaningful. Thank you so much. And I never thought about him having a beard because it was not something I would be aware of. I saw him as a man as my husband and not as a person with a beard or a scar. But as a child he had been burned on his face and had a small scar. He always wore a three-day shadow to hide it but not a beard. That must have been what you were seeing."

This obviously was one of the most poignant moments of her life, and mine—facilitating a needed message from a thirty-year deceased husband to his widowed wife. He sent a healing love note with multiple areas of proof that he continues to exist. He used techniques totally beyond her consciousness, so no one could say I was reading her mind. And all this I was able to see without my eyes being open. My extracorporeal mind could recognize him as if he were flesh and blood being recognized by my intracorporeal mind.

With my extracorporeal mind. I could hear his voice. I could understand his language. I could understand his meaning and once he delivered his message he disappeared. It was not just one thing that the man was able to impart to me. There was the scar, the height, the beard, the watch, the message, the inscription, his age at the time of his death, his last manner of dress and his connection to the host—at least eight specific identification points. When adding these odds to the first two experiences, we have an astronomically impossible chain of events—proof of a miracle happening that evening, in the midst of ten persons. Even if we only give the third vision the same odds of 1000 to 1 the odds of all three happening are 1 billion to 1. That night I personally experienced an event with odds at least a billion to one—an encounter directly with the ethereal mind of the host's deceased husband connecting with my ethereal mind—despite his detachment from this physical world some 30 years earlier.

It appears to me when the physical body dies, the brain and the intracorporeal part of the brain die as well, but the extracorporeal ethereal mind is a replica of everything that is in the physical body and brain, and it does not deteriorate. It is free, it is intact, and as was demonstrated that evening, it remains in the cosmos. Also as demonstrated, it can be as close as our next breath.

To me, this is proof of a caring and intervening God. I believe God intervened purposefully and powerfully for the benefit of all who were there as well as those fortunate enough to read or hear it told now. Naturally, each reader will have their own analysis or "take away" from this miracle. I will present my full analysis in the next chapter.

ANALYSIS OF A MULTI-PURPOSE MIRACLE

As certain as the sun rises daily there is life of the spirit that has our identity (the spirit mind address) and is in no way dependent on flesh and blood to exist.

Please pay close attention to the first glimpse of miracles provided by the events shared in the previous chapter. It is more than a glimpse—more like a session with multiple meanings. An analytic discussion follows:

Discussion Points:

Of note: I would have preferred to decline this invitation but felt I could not offend my patient by refusing the request to attend a spiritual meeting. My personal belief is that it is dangerous to open doors when you do not know what to expect behind them, but I agreed and went.

1. The first vision: My revelation of a significant and joyful aspect of her life as my first vision served several purposes. It was not frightening. To the contrary it was uplifting, pleasant, comforting and anxiety reducing. Much later, reflecting on the full meaning of that evening's events, it occurred to me that we were all being shown that life should be happy, joyful, harmonious and spiritual. The lady excitedly corroborated the accuracy of the vision. She proceeded to add the meaning of it for her family, being happy committed Christians. Perhaps a metaphor to all of us.

2. The second vision: Analysis regarding the second lady receiving the message contradicting her belief of having a secret, namely "No one can know this because my husband is a minister. He would lose his job if the congregation knew." Notice, first I said, "I see sorrow and sadness." Sorrow and sadness can't be seen in an unchanged expression. Obviously, it was not the secret she and her husband thought. The Omniscient Knower signaled, "I know."

3. The third vision: The communication with the hostess' deceased husband was more than she or I, or the group could have expected. My thoughts are abundant as to the meaning, timing, etc. This was during my agnostic but strongly and earnestly seeking years. I had prayed to God,

"If you are truly there and if it is Your will, please show me." And some will argue the first two visions to be a form of mental telepathy. I don't believe so because of the two foregoing messages, it makes more sense in terms of answering my need to be shown, especially in light of the third message. Furthermore, I was allowed to see pain and sorrow that was not known, in a perfectly normal appearing lady. This particularly gave credence that I could see a scar that was not visible, that is to say it was not a guess.

Further discussion:

We humans build, enjoy for ourselves and exhibit to others a public face of our values and character as this session truly did for this group—including me. It is significant that in this first scenario the family is happy bearing the fruit of a joyful, spiritual life and somehow the orchestrator of this message directed its meaning to each of us present either consciously or unconsciously. This message was chosen first as if the joy and happiness of life is something we should consider a priority. The family obviously was living in the spirit of joyfulness, "it's our favorite place." That comes with the true spirit of God as these people, believers in God joining together as spiritual friends, despite different denominations, for the purpose of spiritual guidance. I recognize this as a message to remember the second commandment as a tangential reference to love our neighbor.

On further analysis of the vision with the pastor's wife, the message did not end there. It was evident to her and the others present, that now these other nine knew, and she had to have known the implication of that revelation. I deduce this because she added the phrase, "This must be kept secret." All of us should instantly have been reminded God sees and knows everything. Later I understood, even being confronted openly, admittedly in a spiritual setting she persisted in defying a public admonition. We are often too slow in picking up God's message. It was much later before I realized the message for her was for me, the others and ultimately to everyone, even to our current readers: "I know" meant "I Am the Omniscient Knower."

In evaluating the three cases the images became clearer as the meditation session progressed. The entire drama coincidentally played out start to finish between my first having an inkling of seeing the pavilion and the ending of the communication with the deceased. As if preplanned timing, the session ended shortly after the gentleman nodded and disappeared.

The interaction with the deceased husband proceeded comfortably, as if communicating with a person in the flesh. We mutually understood each other as if we had been physically together. Yet, we were simply looking at each other eye to eye and nodding, with complete and total understanding. At the end, however, when words were needed, I understood his words as clearly and distinctly as I would from anyone, yet his lips never moved.

For example, further details: At the beginning he remained unmoving except for his eyes as he looked me in the eye. Yet, he distinctly appeared to be alive in the sense that motion could be sensed and conversation coming from him was fluid. He was unwavering until he understood that I had appropriately perceived his beard and was seeking to find a scar. He, in charge of the conversation, nodded and proceeded to produce the pocket watch. He lifted it to face level, opened it and pointed to the inside of the lid. I never saw the inscription, but I knew one was there, that it was meaningful and that it should be mentioned. Perceiving my understanding, he returned the watch to his pocket.

To my knowledge, I was not actually nodding my head but mentally indicating a nod. His moving to a new subject signaled me that he understood me and that he was continuing his message.

At the end he stared me directly in the eye and without his lips moving spoke in plain English. When I acknowledged hearing his words, and that I would relay them, with a nod, he disappeared. How does one see something that eyes cannot possibly see? That was the case in all three scenarios, even seeing the sense of sorrow.

What message or messages might we glean from the foregoing? Though there are numerous examples from various people that are similar with my experience, the percentage of this occurring to any one person is quite low; that it would occur multiple times in exquisite detail are odds seemingly incalculable. Why?

If these occurrences were commonly ubiquitous, we could scarcely imagine the scope of chaos to humans if we had unbridled access to the spirit world. It seems to me the spirit world could not be in a "heavenly state" if having to constantly contend with the human world.

The general separation seems to be appropriate and proper, yet occasionally there are dramatic examples of glimpsing the spiritual side. These are special gifts for special reasons. Certainly, sometimes for specific answers, sometimes to have one realize the reality of the living soul.

Given the fact that the widow commented "I have waited decades for this." signaled to me that she had at least prayed for answers or relief from an agonizing past event. Personally, I find it difficult to think of a more convincing answer to her prayer.

I have since often wondered what the impact of that evening of spiritual revelation had been for the others that were there. Did they analyze it, as I had, and did they arrive at similar insights? I remain curious. Some of the group were younger than me. My curiosity begs their input. If any of them are currently still living and happen to read this, my hope is that they will relay to me their recollection of that evening and its meaning to them. Perhaps family members or relatives might have heard these stories; it would be enlightening to hear how they perceived the evening's events and the stories related. To this day, I would be very appreciative if anyone there would be willing to share with me.

For me it was dramatic and profound. I had asked God if He was real, to please show me. This was one of two known prayers answered that evening. Thank you, God.

The cumulative multiple odds at each step of the preceding events of this happening to someone, especially me, still an agnostic would seem to be insurmountable. On the other hand, genius is beyond human. Only a God could stream together so many messages in such dramatic fashion in so short a time in the presence of multiple witnesses. It was truly a glimpse of the ultimate author displaying talent, genius, and power to use communicative skills not measurable by human instruments. At the same time giving specific answers and needed knowledge to a group of people only asking in meditation, 'what is there for us?' God showed divine understanding, compassion, grace, interaction, intervention, and love of ten human beings, enough to plan and orchestrate this intricate play, performed perfectly without actors or rehearsal. The messages cannot be lost, even now pushing the pen that makes it available to the current reader.

PREFACE

Dear Reader, as you begin this book you will encounter a setting unfamiliar to most, even unimaginable by many. It was, and is, real, as is everything in this book, for it is nonfiction.[2]

The reason for starting with this setting is to show you the humility and vicissitudes of my early life. My hope is that this humility will be evident throughout the book and remain at the end of my life.

This book requires being read in the sequence it is written. Skipping to a chapter with an appealing title will cause misinterpretations—for not knowing what transpired before.

We derive meaning from our lives both with our five senses (human/corporally), and with what we glean of unseen, extracorporeal blessings and interventions. We may or may not be conscious of the latter, but we should endeavor to develop

[2] Where examples are fictitious, it will be noted.

awareness of them if we hope to know ourselves fully in anticipation of our ultimate destiny.

What we are able to measure includes our Milky Way galaxy of 200 billion stars[3], with whatever planets are around them. Our one galaxy resides in a universe of 200 billion galaxies.[4] For all we know, there may be other universes, however I personally do not believe there are. But, of our universe all the stars and planets comprise only 4.9% of all the cosmic energy and matter that is total and measurable. Scientists know this 4.9% to be real and present to our five senses. Incredibly 95.1%, although calculable, is not sensibly or materially tangible to us.

Many will be surprised to learn the fact that each of us incorporates a portion of dark matter, (energy matter) from outside our sensible universe. Interestingly, it plays a significant role in our lives and is the only part of us that cannot be destroyed.

Our five senses are designed such that we can only experience things of the 4.9% because we are created in that intentional design. However, God allows exceptions. The other 95.1% is aware of us and intervenes at will in our lives. These interventions are often what we call miracles. People (angels) appear to help and then disappear. We see them in miracles, but we cannot capture

---

[3] Estimates vary greatly.

[4] Estimates vary greatly.

them with our scientific instruments. Sharing my journey and subsequent conclusions are informative and a partial reason to write this book. It elucidates knowing we have a creator and shows how we can—and do—witness His continued works on our behalf. Abundant proof can be recognized and observed. God is alive and active in our lives.

Truth demands that both theology and science require compatibility. My intention is that this humble offering shows that reality.

This book is not a life biography. It is a mixture of scientific fact as currently conceived, and spiritual facts. I also share some logical, biological pros and cons for doubting or disbelieving, as well as the same in the case for belief. I will share how all of these have affected my spiritual journey and that of others, whether they recognize it or not. Details are revealed throughout the journey. Many might be surprised to find that the use of logic is key in the spiritual realm as well as in the scientific. I invite you to follow along with an open mind. My prayer is that you find something in these pages that will further you along in your own spiritual journey. Your reward will be to find joy with each dispelled doubt, and blissful freedom, relief and peace in your belief. Those who are already believers through faith should realize an added comfort. Doubters should recognize how tremendous the odds are in the favor of belief, and realize that they are too overwhelming to wager one's soul against.

If any reader is, as I was, a doubting Thomas, please consider the mathematical odds as you read. Only then decide. Ask yourself, could all of this—or indeed any of it—have happened simply coincidentally? When we see something unexplainable in our world we ascribe super normal forces to be at work. Often these events are miracles. The 40 plus miracles shared here should help elucidate and give meaning to God's intervention in our lives, including miracles.

INTRODUCTION

Scripture is Science – Science is Scripture

This is about compatibility between the two, and the intimate relationship between God and humans, made in His image.

It should be read sequentially. Not to do so would in some chapters severely challenge the progressive nature of the information causing disruptive and unintended interpretations.

We cannot know all there is to know about God, or the expanded cosmos, but to ignore logical conjectures based on extrapolation of what we do know is to play and act dumb, much like the servant given one talent as opposed to five or two.[5]

SIS, Scripture is Science – Science is Scripture, suggests we use all our talents through the abundance of spiritual, cosmic and scientific information designed to tether us closer to the God who

---

[5] The parable signaled to us that we should use the talents to the best of our ability until His return.

ESV Matthew 25:30

created us, loves us and continually showers us with His grace, when even we don't feel worthy.

Appreciate it and give Him the glory and thanks.

Miracles are a strong part of this book as they show us His divine interaction and power.

Passion for truth in the sciences and religion generally references our point of origin, i.e., creation. We question, are we and the universe God's handiwork or just incidental to "the big bang." This point of view misses the mark by trillions of years.

The full perspective indeed shows us how we are created in God's own image by God alone. But it was unaccounted millennia of God's inventiveness and self-education before God gave the word[6] which initiated the beginning of our universe some fourteen plus billion years ago.

We understand from this sequence, our beginning came by way of a mature, powerful, living, thinking, intelligent God. We and our universe are the second greatest accomplishment and story ever to occur (to our knowledge). God, however, is the Greatest accomplishment of all and the total credit, as we will show, is to none other than God!

---

[6] (volition = command)

The solid facts and sound theories in this book indisputably show God as a living being, leaving not one inch in which to doubt God as a creator. God's supremacy and His omnipresence show Him to be a living being, not subject to death, and being willing and able to intervene at will, or not. Many examples will show how scripture and science actually validate each other.

God is now, and never has been, without free will, and He will keep it forever. God gave us free will in a brain with multiple choice options, but God also imbued us with a conscience and a mind dually tuned to God's frequency and to its host brain.

This book validates those who submit to a divine calling by faith. It provides a comfortable and well-worn path for the materialist and the doubting Thomases. We must do two things, stop at each reference point and digest what part pure materialism played and what part divine design played. As we walk this journey, and at each of these experiences stop to calculate the accumulated odds—one after the other; we must ask ourselves, can we win a debate or wager against these odds? Would we "bet the farm?"

Throughout this book I will show you the answers through the use of imagery and stories: stories which contain a number of miracles. Some involve me personally, but there are also stories collected from many others. All of the stories shared are true.

The first two miracles that happened to me you will see early in the book as a way to introduce my background, and to the beginning of the series of my personal miracles and more.

LOGIC FAILED HIM

Joyce, an RN, and I were having a casual conversation when, as often happens when two medical people converse, the conversation veered to particular experiences of interest. On this occasion we were recalling some of the stranger experiences we had seen in our careers. Upon recalling one from years before she exclaimed, "Oh, I have to tell you this one!"

Joyce is a pediatric nurse who was working in the children's unit, and on this occasion she was assigned to a particular infant. To my recollection the infant was less than six months old and not doing well. Symptoms and studies indicated that the problem had to do with the intestinal tract. The only hope as recommended by a reputable pediatric surgeon was to have an exploratory look to see if anything could be done.

During the operation he found the intestinal tract to be severely gangrenous. Knowing that the situation was hopeless, essentially a death sentence, he closed the abdomen and informed the family of the situation. He was sympathetic to their grief.

Following the surgeon's exit, the family decided on a prayer vigil. They included other family members and encouraged them to join them in prayer. Throughout the night they prayed that the child would be healed. During the night the mother felt certain that she had assurance from God that her baby would be okay. They continued their prayer until the surgeon arrived for his morning visit. The baby was still alive but that was about it. The surgeon's news continued as a dire prognosis.

The child's mother informed the surgeon of her message through prayer that God had given her assurance of improvement in her baby. She felt "certain the baby was better this morning".

The surgeon was insistent that with his experience and knowledge there was no logical way that the baby could have improved from what he had seen in the operation, and that they would need to accept it.

The mother stated that she could not accept it because she was certain that God had given her the hope and the knowledge that her son was better. She pleaded with him to reopen the abdomen and take another look.

Reluctantly he relented. He scheduled the surgery. This time the surgery lasted much longer than the first time.

The family noticing the time elapsing thought this had to be good news but of course they remained anxious.

When the surgeon came into the room, he lifted both hands as if not knowing what to say, but his words were tantamount to "I've seen a miracle, you have a miracle son." He told them that a significant amount of the intestine had indeed pinked up considerably. Some of the intestine could not be saved but enough of it was saved that he should have an adequate amount of intestine to survive normally.

This nurse is someone I know well; I trust her story. It is even more significant that she lives in the same community as this family, and their children go to the same school. It had only been a short time before our conversation that at a recent school function she had seen the boy, then seventeen years old, in perfect health. She has followed him through the years and watched him grow as normally as any of the other children.

Her expressions relating the story signaled the impact it had on her. One can only imagine what impact it must have had on the boy's entire family, and hopefully the surgeon as well. These strong miracles are shared with hope, and prayer for expanded impact.

This miracle occurred in a town and hospital twenty-five miles southwest of Birmingham, Alabama.

FOREWORD

This book is another revelation of the reality of God. It clearly shows His intimate caring for us. You will encounter approximately three dozen miraculous such episodes.

The author believes a mainstay of God's keeping our attention, both in Biblical and present times, has been through miracles. Most adults have some knowledge of Biblical miracles.

The miracles from God continue today. As you read this collection of stories, you should have no difficulty realizing their miraculous nature. Many present-day people have a personal miracle story. Most consider this a special gift and are happy to share. In fact, that is how in this book you will find dozens of shared miracles.

You might wonder why you have not had a personal miracle, if indeed you have not. Some, as they retrospect over their life, uncover what they believe was a miracle in their past, but perhaps was not previously recognized as such. While the book shares approximately forty miracles, there is one intriguing bit of information I choose to share here at the beginning.

Everything in our universe is made of energy/matter in a circumscribed wave/band width. There is one exception. Humans—and we have rational reason to believe, most larger animals including dogs and cats—have a mind separate from the brain.

At some level of the descending animal scale, the brain is sufficient and no "mind" is needed. The mind is composed strictly of dark matter, this being the energy/matter of God and His Domain. It is genius on God's part. Things of our universe are destructible. Dark matter is indestructible. We are in effect one hundred percent preserved by way of our mind as it perpetually records and preserves everything about our lives. It doubles as our soul and therefore is liberated at death to its true home, the dark matter.

THE EARLY SETTING

The setting is a two-room sharecropper's house. A husband, wife and an extremely ill, very young child are inside. There is no running water or indoor facilities. Water must be drawn from a well via a bucket, a rope, a spinning pulley wheel. Toileting must be at the "outhouse." The year is 1937. It is a mid-February cool evening. The parents anxiously await the country doctor's arrival. Wood is burning in an open fireplace.

The area is sparsely populated. One and three-quarters mile to the east is a crossroads. There is a hand operated gas pump, a modest grocery store, and three houses close by. Four miles west is Harrison, Georgia, population approximately 250. North at approximately nine miles, Tennille, population approximately 1250. At approximately eight miles south, Wrightsville, population approximately 4900. The nearest hospital was in Sandersville, population approximately 10,000, is three miles beyond Tennille.

East at roughly one-half mile, on a side road to the left at one-quarter mile is a more substantial house. It is owned by the country physician, Dr. Thomas E. Vickers, who also is the landlord of the sharecropper's house.

Inside the house, the baby's mother Mae, is twenty-five years old and has a fifth-grade education. In addition to the regular household chores and duties she helps tend the garden. Vegetables are grown in the spring and summer. She will harvest, cook the meals, and can the preserves for the winter. During blackberry season she makes blackberry wine.

Currently, however, much of her time was taken with her young son, just twenty-six months old.

Horace, the baby's father, is thirty years old and has a third-grade education. He is a sharecropper. His responsibility is to produce a crop for the landlord and his family.

In this unimpressive setting, the anxiety and concern of this meagerly educated couple for their extremely sick child is made worse because their firstborn, Allen, just two years earlier had been born with a lung affliction and had lived only two weeks.

Dr. Vickers drove into the dirt yard. At the door he rapped twice with his right fist and without waiting, opened it in his usual manner, stepped inside and closed the door behind him. He sat his large well-worn medical bag on the straight-back chair positioned strategically for that purpose and so that he could hang his topcoat, which he did. Walking toward the crib he asked, "How is the boy?" Hearing "No better." he began his examination with the abdomen, the area of most concern. Next he listened to the heart and the lungs and checked the baby's strength and weak reflexes.

Mae and Horace noticed he was lingering longer. Their anxiety soared. The pungent smell of asphidity permeated the room. (Sick children had to wear a necklace laced with it because of it being so offensive, no child would come close.) Mercifully, the smell would leave with Dr. Vickers' bag.

Finally, he raised up, turned and walked toward them. Replacing the stethoscope in his bag and putting on his coat, he informed them he was saddened to give "a bad report." He said he knew that for them it was even worse because they lost their first child, but they were about to lose their second child. He would not make it through the night. He said he was sorry. They began to cry. He took a couple of steps toward the door, opened it, turned back, and said again that he was sorry, but that it was now "in the Lord's hands." With that he exited, closing the door behind him. Already sobbing, the parents barely heard the car start and leave.

What this modest house was hosting, and this couple would experience next could be said to have predicted this book. I have no proof of that. The beginning could have been at the birth or death of their firstborn. Who knows how long it might have been planned without human knowledge, or even if it was? Each reader will need to judge for themselves, as indeed have I, but not in a straight trajectory. We often believe first then search for scientific proof of God's foreknowledge.

What Mae and Horace would experience played out spontaneously, but they could not foresee the twenty-five years of consequences awaiting them. If so, would they have opted to take a different path? What on earth—or perhaps beyond—caused this fateful night?

With Dr. Vickers' exit both immediately went to their child, stroked the moribund body, cried and prayed for a few hours, but progressively the infant only became worse. Finally, one of them would remember the doctor's words which they discussed; could it be that their child's fate truly could be in the Lord's hands? Should they be praying directly to the Lord, specifically for healing? They thought they were. It occurred to them they were praying for something miraculous but offering nothing in return. What could they offer? Having essentially nothing, they agreed to raise me, that infant, as a servant to the Lord, envisioning that I would become a missionary. That night, they offered me, (and themselves for the effort it would bring upon them) to a life of service to God. According to their much later recounting this to me they stated at that very moment I began to improve and by morning was in a much better viable appearing state.

My parents never told me what their spiritual life had been before that incident, but I can tell you truly from that time on they lived their life in the service of their Master, Jesus Christ.

Would an agnostic think such a barter to be unfair, or even unchristian? I could not understand or speak then: but perhaps later. Besides, God's grace is free, right? Not bought with works or money. But then God does make covenants or contracts, doesn't he?

Postscript to this chapter:

The foregoing I have considered to be miracle number two in my life.

Miracle number one was my birth: weighing in at 12 pounds and 13 ounces.

Remarkably, it was a home delivery by Dr. Vickers. The only adverse sequelae to me from my birth was a mild asymmetry in my growth, with my right side winding up ¾ of an inch longer than the left.

I was curious and fact based at a young age. At four or five years old, I remember approaching my mother with a marble in a sack, and then closing the top tightly and putting it in front of her and exclaiming "I know God cannot see the marble in this sack." They were in fact raising an inquisitive child who would be observant of the contradictions so obvious in professed religion with contradictory lifestyles.

My curiosity was really a need for proof. It parallels many, if not most, of the people who are introduced to the concept of God. The search is particularly bothersome when expert theologians question science and science disputes theologians. Most of us are caught in the middle, without the knowledge or understanding to reconcile the two.

I do consider myself fortunate for having chosen the path of medicine as it required the study of the sciences of biology and physics. I learned enough to gain appreciation and knowledge of the sciences, and to be able to apply them in my day to day thinking and their relativity to life and matter and cosmological realities. Therefore, both science and religion (or spirituality) have been a driving force to me, to find whether either or both are true, and whether or not one is ruler over the other.

My journey is everyone's journey. Some of my experiences might be more dramatic than many of those who read this book. Likewise, some readers possibly have experiences even more dramatic than mine.

We should all continually remind ourselves to stay vigilant, keeping our eyes, ears and mind open for all things scientific and all things spiritual.

A TIMELY VISIT

Robbie at 63 mentioned recently having ponderous thoughts regarding his life. He reasoned that it might be because he soon would reach his sixty-fifth year. It seemed close to the magic number, "three score and ten" or 70 years of age. Nevertheless, he was thinking of the good deeds he had done; but realizing he was not perfect. He was weighing the positives against the negatives.

Fortunately, Robbie was able to assess himself with more positives than negatives, giving him a reasonably comfortable feeling about himself. He then pivoted the conversation.

Robbie recalled an incident that happened when he was forty-two years old that changed his life dramatically. He credits, with appreciation, this incident as having shaped the remainder of his years, and his resulting satisfaction with the changes he had made. His story is interesting.

One day no different than any other, he left work to go home. But when he arrived at his home a strange event awaited him. He let himself in, as his wife was not yet home. He walked directly to the bedroom where he intended to change his clothes and relax. When he stepped into the bedroom, he saw a man sitting on the bed with his back to him. At first, he thought this was a friend. Looking more intently, he realized this was not a person, but a spirit. The entire figure, clothes and all, were glowing white. He was startled. He remembers it as distinctly as if it were yesterday. A question to the spirit revealed his thought, "Spirit, have you come for me?" There was no reply. He repeated the question with urgency. "Spirit, have you come for me?" Still, there was no answer.

Robbie next said "Spirit, if you have nothing to say to me, I am leaving, because I am scared."

He went to his car and left. Quite shaken, while driving he contemplated why the spirit had shown himself and yet said nothing. This incident was the first time he had truly contemplated his life, the meaning of life, and weighed his own 'pluses' and 'minuses.' That original assessment was not as positive or satisfying as this more recent one. He surmised that the spirit came to remind him that though he could have taken him, he did not. And that instead he wanted him to realize that the next world holds the true meaning of life, and that we should use this life to prepare ourselves for that time and place.

Robbie shared that this had been a profound and life changing experience. Although he had been frightened, he nevertheless received it with a personal satisfaction of caring and a new direction which he maintained throughout the last twenty-one years.

I have known Robbie for many years. I know him to be honest, truthful, reliable and trustworthy. He, like many, had a very intimate encounter.

Since everyone does not have a personal encounter, it often gives satisfaction to a believer who can resonate with and feel the miraculous power through identifying with stories from others. They could come to us in various books from different angles, and even at times in movies.

A LATER LARGER SETTING

My childhood was one of being a magnet for illnesses and mishaps. I had a three-year younger sister who was generally the healthier of us. The one constant in our family was persistent financial difficulty.

But then, despite the dire financial situation, something shockingly strange happened. In 1951 a shipment of a complete set of Encyclopedia Britannica arrived at our home. It was gorgeous. It had cost $400.00. My father bought the set for my sister and me, believing it would enhance our education, despite its outrageous cost for him at the time. Little did he know just how that purchase would change the trajectory of my life.

At that time, my impression of the world had been limited. We had truly little radio, no TV and I had had no encouragement to read anything outside of comic books. Perusing these beautiful new volumes seemed to me that it housed as much information as the Library of Congress. I was hooked on knowledge and on learning as much as possible from that moment on.

Imagine my fascination, as an impoverished teenager with little education, learning that there were literally entirely new worlds of information available! Pictures of rockets drew my attention. I read of Werner Von Braun, and of aspirations of moon visits. I used the information I learned to calculate that to my estimation, to make a reasonable trip to the moon for a few days and then back would take about two weeks. I read about Winston Churchill. In reading, I learned about our awesome planet. I read the entire set, cover to cover!

By then we were living in Tennille, Georgia, still within ten miles of my birthplace, but this new knowledge had given me a much broader world view: one that was not only global but cosmic. This taste of education changed my life forever.

For the first time I found reading to be exciting. Outside of comic books, homework had not been all that exciting. This new knowledge, along with my imagination, allowed me to break through the previous geographical boundaries I felt were constraining me. Einstein and Werner von Braun became my idols. I latched on to science as if it were the ultimate.

Being young I was certainly impressionable. Einstein stated that the fastest speed possible was the speed of light. I believed him wholeheartedly. When he doubted quantum physics, I was right there with him. In college, pre-med was very heavy in basic science, but unfortunately, between classes and working three part time jobs, my schedule left little time for pleasure reading. Later, in medical school the schedule was even more demanding. There was little time to keep up with scientific progress, but the bits and pieces that I heard here and there seemed to indicate that science was progressing, so I remained enthralled and comforted by it.

One thing that puzzled me was that Einstein believed that the universe in its entirety was so precisely functional that it essentially demands that there be a creator, but he could not get comfortable with the creator being of a personal nature. This was a stumbling block for me. When one has an embedded strong belief system, anything that challenges that belief causes anxiety, and the natural tendency is to fight the change. This has a rational scientific basis called normalcy bias, discussed later.

Latching on to the wonders of science, traveling to the moon, and finding distant galaxies was thrilling to me—enough to let my imagination escape nonscientific beliefs. I figured if it takes 8 minutes for light to reach us from the sun and it is the fastest thing, there is no way there can be any kind of being that can be involved in something as large as a galaxy, let alone a universe. I was gathering information, I had some answers—though at the time, not adequate information or answers. However, I had learned enough to create a disconnect with the faith I had been raised with. To me, this is an example of the axiom that a little knowledge can be more harmful than no knowledge. (Discussed further in the next chapter.)

Eventually scientists came to the discovery that quantum speed is so fast that the only word they can use for it is instantaneous. As far out as they had been able to measure it, it is immeasurably still instant! When two electrons are spinning and parallel, and you stop one, or change the course of one, if it is paired the other instantly copies exactly the change no matter how far apart or how distant. This is one way to conceptualize quantum speed.

Amazingly, science discovered that by the pure process of human will the wavelength of energy can be changed to an electron or the electron can be changed back to a wavelength. This is proof of the "word" being tantamount to God. If God is the major energizing force His word is His volition. Therefore, if He desires something, or wishes something, or speaks something He has command of the energy available to make it happen. We do the same thing to a lesser degree. There are many variations of why some find this easy to believe and some do not, and the variations on this point are profound. Depending on a person's normalcy bias it may be easy to accept or extremely difficult to accept. This is explained later.

The new setting is cosmic. We must move away from our earthbound beginnings and nurture our unbound scientific and spiritual potential.

GOD IN OUR LIVES

Early in life we cannot see how our life will unfold. Later in life looking back, it's as if we have a panopticon perch where we may see everything—and how seemingly unrelated incidents were woven together for maximum effect. We are lucky and fortunate if we become consciously aware of this, because we will better understand ourselves and our lives.

Looking back in this way, the following two incidents proved to be more related to a couple of incidents discussed later than I could ever have imagined at the time. In my third year of college at Emory University in Atlanta, Georgia, I lived in a boarding house a full block from the campus. There were fellow students who lived there as well, and having no car at the time, we all needed to walk to campus, and then to our respective buildings for classes.

One particular morning, another student and I—loaded with books, notes, etc.—waited on the front porch for a break in what had been an unrelenting heavy rain. We became less and less hopeful there would be a break in the rain's intensity. Our anxiety grew as class time drew near.

As a little background information, I had stayed out of school for the previous fifteen months to pay off some family debts, and during that time I had become a magician, and had performed some stage magic and hypnosis. This fellow student was aware of that, so on a lark, I said, "Hey, I'll run upstairs and get my magic wand." We both laughed, however, I did go in and get my magic wand. On returning, I removed it from its case, pointed it to the cloudy raining sky and—with words I don't exactly recall—demanded the cosmic powers to stop the rain. Amazingly it happened immediately. He was profoundly impressed. I was shocked. At the time, I considered it pure coincidence. Practically speaking, I gave it no more thought. It never occurred to me that it would be bookmarked in the unabridged library of the cosmos. At that time, I didn't even know if there was, or could even be, such a thing.

Two or three decades later, something happened that brought back the memory of that incident. I had recently read Jonathan Livingston Seagull, and was thinking about the book while enjoying a walk on the beach. There had been a mention in the book about wishing for a cloud to disappear. I looked to the sky and there were several white clouds floating. I remembered that morning years earlier when my "command" had stopped the rain. Something in that book had brought up that previous, long forgotten memory.

When I looked up at the clouds I thought, could it be that there might be some connection between our will and the clouds, or to be more specific, between the Creator and our desires? I wondered if it was heresy to even think that way. I wanted to test it, but I didn't think I should. Finally, I decided I would. I studied the clouds. There were various shapes and sizes, and a significant number of them—a perfect environment for my "test." I chose one that was smaller than medium size, but larger than the smaller size; a significant white cloud. I asked God if He would make that cloud simply go away—leaving the others intact.

Uncanny as it was, I was not even surprised when the cloud began to disappear and in short order it was completely gone. All the other clouds remained the same until well after the selected cloud had completely disappeared. After all, I was not a full-fledged atheist, rather what you would call a 'Doubting Thomas.' I remember thanking God and adding that to a list of events that helped me along the way to believe that not only is God there, but He is truly with us.

On the occasion when my mother died, I lived in a city distant from her. After the funeral, on my drive home I looked at the clouds. I clearly saw the profile of a lady that could be my mother lying with her head on a pillow in a peaceful repose. This vision comforted me and my sister when I related it to her later.

We will come back to these, as they are a prelude to God showing up twice more by way of clouds, the latter of which convinces me that stopping the rain that many years ago was indeed a bookmark event in my life. One that God would recall to me decades later in one of the most poignant experiences of my life.

INTRODUCING SCIENTISTS

I believe an apt description of "scientists" would be: researchers and discoverers of the laws of God our Creator. During much of my younger life, my concept of a scientist was limited strictly to our civilization. I had no concept of prior civilizations having scientists. I'd particularly like to mention three prior civilizations: Atlantis, Lemuria and Mu.

Most people have heard of Atlantis. Perhaps they haven't heard of Lemuria and Mu. Some historians think Mu and Lemuria might have been the same, some think they were more like neighbors, while still others think they were separate civilizations. The inhabitants of Mu are known for having had highly advanced scientists as well as a strong spiritual culture.

Astoundingly, this was ten thousand years ago.

Information through the descendants of Mu and the study of Lemuria teaches us, not only through their scientists then, but as well as our scientists now, continuing discoveries of the laws of nature set in motion by our Creator.

Relics from Lemuria include a sophisticated calculating machine, evidencing their inventiveness.

The next chapter is a fascinating story reaching all the way back to the civilization of Mu, 10,000 years ago.

STEALTH SCIENTISTS

Giza, Egypt, circa 4500 BC:

A group of nomadic people roamed the countryside of Egypt. They traditionally settled in communities for short stays. Occasionally they would leave a member or two behind when the group departed. We know them as Hyksos. They were the descendants of the few survivors of the latest destructive cataclysmic topographical earth makeover many centuries before ours.

The Hyksos were different: characteristically kind, gentle, intelligent, and extremely knowledgeable—including advanced scientific knowledge. They entered villages and gladly shared information such as tool making, when and how to plant and tend crops, when to harvest, and other useful techniques for better living. Because of this, they were immediately accepted, honored, given high positions and revered as intelligent teachers. Secretly, they held knowledge of scientific facts and prophetic information that would be useless to the then current civilization.

As you read this chapter, most will see the Creator, sustaining God, His religion and science as one cohesive inseparable unit, as have I. We are even more blessed to be able to feel, understand, and enjoy that privilege.

A large contingent of the Hyksos settled in Gaza, Egypt. After establishing themselves in the community, they proposed building a great pyramid. Describing how magnificent and wonderful it would be, they presented a detailed project plan to the rulers and community leaders offering to oversee the project. They had no difficulty enticing the labor forces to build it. Today it is known as the great pyramid of Giza, standing approximately 40 stories tall – 481 feet. It required eighty years to build, being restricted by the rainy season. It would remain the tallest man-built structure on the planet for over 40 centuries until in the 19th century when a structure eclipsed the record.

The prophecies and scientific facts built into the pyramid were without words, only scientific markings. The scale used by the builders was not discovered until the 19th century. It was discovered by dividing the diameter of the earth by 500 million resulting in a 1.0011 figure, called the pyramid inch. Discovering that scale was like a key that unlocked centuries of secrets held by the pyramid; not only facts, but prophecies as well that had been hidden within.

For example, a straight line from one corner to the other equaled exactly our calendar year. It was not built in a straight line. The base was built in an inward and upward curve giving two additional measurements. All three of these measurements exactly match current scientific measurements of our year using three different techniques to measure it. The base of the pyramid covers thirteen acres. They based all interpretations on a pyramid timeline discovered by researchers to be 4000 BC to 2000 AD.

Significant prophecies hidden within have stunned archeologists, scholars, and scientists. The following in particular: World War I was declared on the day prophesied, as was the precise day the Armistice was declared. A skylight pointed to where the Star of David would be at the time Christ was born. A further prophecy noted the exact dates of Christ's birth and crucifixion.

This is incredible information. God used the Hyksos, harboring scientific knowledge, to signal us 6000 plus years later. He already knew those events in detail. Even before our civilization this previous civilization had advanced scientists who passed their knowledge to descendants. God used them to inscribe His message in stone of the coming event of Christ's life and death. We only discovered the message in the 1800's AD.

Much of this information can be found on the internet or in various books. I like *"The Great Pyramid: Man's Monument to Man"* by Tom Valentine, copyright Pinnacle Book, Inc.

In the mid 1800's C. P. Smyth, a scholar and man of conviction regarding the authenticity of the Divine message gleaned from the pyramid, was receiving a great deal of criticism. Smyth smartly sought out David Davidson, a respected scholar, structural engineer by profession and trade, mathematician, historian, architect and astronomer. He was an openly strong agnostic. He would investigate and evaluate the pyramid and its alleged Divine messages. If any person could prove the pyramid to be void of divinity, he would be the one.

Davidson agreed and began his examination and evaluation. He spent 25 years measuring, remeasuring, digging deeper and deeper into details. Finally, he acknowledged the Divine imprint as being overwhelmingly compelling and undeniable. He became a Christian and co-authored an extensive book, *"The Great Pyramid: Its Divine Message"*[7]. It was said to be more extensive than any other pyramid book, with more information in greater detail.

Here we see a divine imprint where persons I am calling stealth scientists worked with ordinary unsophisticated laborers, yielding the most magnificent structure on earth. It has stood proudly now for 6500 years. This shows the incredible ability of God's forecasting the future in such detail and His power to insert information into the human psyche to do His bidding.

---

[7] *The Great Pyramid*: Its Divine Message, Kessinger Publishing, LLC

The pyramid is extremely specific in drawing strong attention to the life and death of Christ. The skylight pointed from the king's chamber to the Star of David, exactly where it would be at the birth of Christ. Mentioning the birth and crucifixion of Christ was incredibly exact.

Curiously, the year 1844 was mentioned without specifics, therefore ambiguous and confusing. In that year religious events were profound and global; the consequences were seismic. 1844 encompassed a multi-denominational religious fervor. A mighty global urgent sense of spirituality leaped forward that year. It arguably would rival any global spiritual awakening since the days of Christ and his crucifixion.

Many years, more than a century now, have passed. Time has drained the fervor of that moment and robbed many memories, greatly diminishing the retelling of the 1844 phenomenon. Thankfully, much was recorded for those seeking historical spiritual truths. Here are some highlights.

A minister by the name of Miller was at the forefront of a belief that Christ would return that year. The belief became international. The estimated number of active believers ranged from 100,000 to 500,000. A large contingent was in the United States, whose population at that time was around 17,000,000.

From around the world many believers from diverse religious backgrounds gathered at Mount Carmel, often dressed in sackcloth and ashes, waiting for the expected return of Christ. The Seventh Day Adventist, still in their formative period, were present. They seemingly believed the timing of their formation along with their being present was a sign they would be receiving the truth. Members of the newly formed Baha'i faith were present. Present also were believers from various denominations whose identities seem to be lost. The entire event viewed broadly, showed a spiritual resurgence during 1844.

Fortunately, there is a great deal of history about that date but the memory and attention is waning significantly. It became known as the year of great disappointment. Those events turned out to be important to the formation of the Seventh Day Adventists who, after great study, proclaimed that the date was correct. They reckoned it was the expected meaning that had been misinterpreted. The Seventh Day Adventists believed they would fulfill spreading their message of truth, believing truth would be revealed to them.

Several scriptures seemed to point to this year. One was attributed to Christ speaking about the gospel being spread to the entire world before the end time. 1844 was generally believed to be the year of that accomplishment.

Mohammed prophesied there would be no further messenger until iron beats against iron and is heard in distant places. The telegraph had just been invented. Its first transcontinental message read "what hath God wrought". A U.S. congressman introduced an HR bill to close the patent office reasoning that the telegraph was such a far-reaching development that surely nothing would surpass it.

Each religious entity seemed to be more energized in their belief system during the year 1844. But the year 1844 had further surprises, adding more intrigue to the year of ambiguity.

Friedrich W. Nietzsche of Prussia was born 15 October 1844. He was a brilliant man, becoming Chair of a prestigious department of philology at age 24. However, he placed his brilliance and beliefs above religion or secular morals. He did not believe in God. For the remainder of the century, he was an antagonist to all things religious. He perceived those believing in God to be inferior. Perhaps he was a forerunner to the antichrist. He died on the 25th August, 1900. He likely never knew the irony of the timing of his birth and life.

God used the pyramid and Hyksos to prophesy good and evil juxtaposed in four numbers – 1844. The profundity of this monumentally divine miracle is intellectually staggering, as is painstakingly documented by David Davidson's 25 year long attempt to disprove its connection to God. He had no rebuttal to the pyramid's logical messages, nor could anyone else have.

God gave us messages through the prophets, Christ, the disciples, the Gospel and the Holy Spirit. However, the original doubting Thomas needed more. He obtained physical evidence by way of

Jesus appearing to Him. What might we look to for satisfaction of our physical or scientific doubts? We have the answer in the great pyramid. The back story is fascinating.

God's existence and foreknowledge is proved scientifically in the great pyramid. God knew that the doubts of Thomas foretold of a large population of doubters. Many of these doubters would be bright and knowledgeable, David Davidson being a perfect example. Conclusive scientific proof would need to be available. To answer this need, God set in motion a truly complex miracle. The complexity and exactness of any necessary evidence would need to be so extraordinary, and so to the point, that we must concede that it could only have been foreseen, planned and executed by a true God. The pyramid offers exactly that type of proof, and much more.

Consider this: without knowing the facts of the pyramid, not one of us would have sat down and contemplated the following.

God began this miracle by giving these prophecies to scientists of a civilization so distant from ours that what little we know about them is from the memories of their survivors, called Hyksos. God entrusted His prophecy to only these survivors, who would not allow their memories to change over time as is the usual fate of messages passed through multiple generations. If 10,000 years is correct, that means that these amazing people carried the exact details from generation to generation, for 5,500 years before the building of the pyramid in 4,500 BC , totaling 10,000 years BC, thus proving that God knew at least that far in advance the dates of Christ birth and crucifixion.

This group would propagate and stay alive, until their divine destiny, for centuries and know they had a mission to build a pyramid. They would need to find the right place, time and culture, with adequate water, stone, minerals and indigenous cooperation. The structure would require withstanding weather, looters of façade pieces, as well as desecration. Despite all adversities God's message would not be harmed. God used scientists from centuries before to speak to present day scientists, and to all who have doubted, or might doubt still.

This last page is theoretical, logical, deductive reasoning based on the facts before us and our concept of God. How could anyone but a sovereign God of the universe know with detailed accuracy when Christ would be born and crucified or what would be the happening in 1844, 6,000 years in the future, or exact dates and circumstances over a six-thousand-year period, plus whatever time the Hyksos had wandered following the destruction and vanishing of their civilization. It obviously had to have been many centuries since no physical structural evidence remains from them. David Davidson was right: the evidence is "overwhelming and undeniable."

Further Thoughts and Consideration:

The foregoing chapter partially unravels a mystery of a pyramid built four thousand five hundred years ago. I try to understand how the scientific spiritual culture of Mu, extinct more than five thousand years prior to their descendants building the pyramid, came upon this knowledge. Did they have a David-like dreamer, a Nostradamus or a prophet whose name is lost to history? It is curious that scientists would have picked two linked dates so far in the future to be worthy of mention and preserved as important enough to be "cast in stone." It does not seem rational or reasonable that science alone could have had any way of choosing those two dates, the birth and crucifixion of Christ. They would have had no way of scientifically projecting a Messiah that far in the future except through prophets or seers, mystics. Nonetheless, being a highly spiritual society, it was ingrained enough and important enough to God to make sure it was inscribed. It seems to me that directive so far in advance and also included a skylight pointing toward the Star of David would be our Creator signaling His message to the world that the birth and crucifixion of His son would be as prophesied and be the single most eventful occurrence in the human experience.

God further underscored this with the founder of the first monotheistic recorded religion, Zoroaster, who prophesied the birth of Christ signaled by a star. Although it is known knowledge, if not ubiquitous, the three wise men who found Christ were Zoroastrians following his prophecy.

To an oddsmaker accepting the facts outlined here it should be easy to accept a creator and Christ with His message regarding the Father.

Yet, having learned this knowledge early in life, still we struggle for ways to find proof. Miracles, especially personal miracles, are one such way that God touches us with His supernatural abilities through grace and compassion as a bright light on our path.

PS: In full disclosure, I have this past week come across another article claiming more religious figures mentioned in the pyramid, Zoroaster and others, but no author, no source and no references. I must defer to potential future endeavors.

TRENDING SCIENCE DOUBT vs. FAITH

I have always had an inquisitive nature and a highly developed sense of curiosity. Even as a child, I would always ask "who, what, when, where, how and why."

One specific incident comes to mind. In my preschool years there was no such thing as air conditioning or central heat for most people, and no electric cookware or electric irons.

I was reasonably observant, noticing for example when my mother needed to iron something, she would place the flat iron in front of the open fire or on the wood burning stove until it was warm enough to iron the garment. The first time I witnessed her using an electric iron I wanted to know if this was an iron; and how it worked. She explained that the heat came from the cord with the electricity and that it was hot, and I should not touch it. It did not seem reasonable to me that such a thing could be possible, so naturally, when she turned her back, I touched it. She was right.

My parents had me in church every time there was something I could attend. I learned Bible verses and heard the stories of the

Bible regularly while growing up. When I was eleven years old, I joined the church. As I was walking down the aisle, the town alarm system began blaring. I thought that to be unique, maybe even appropriate. My thoughts were something to the effect of "this is something to remember the importance of this moment by." Perhaps akin to marking the moment with significance, or some sort of Divine acknowledgment, of course without knowing those exact words at that time.

Later that day I learned that the sirens were to announce the end of World War II. Even later I learned that it was a false alarm. Much later I realized that though my sentiments were genuine on that Sunday morning, the lack of depth in my understanding belied the fullness of meaning. The act had been intended to be full, nevertheless it was not complete. In some respects, I came to realize that like the alarm, it also had been false.

During high school I began to notice the contradictions of people's words on Sunday and their actions Monday through Saturday, furthering my disillusion. My doubting kindled with time.

In my third year of college, I took a course in comparative religion. The first day of class the professor dramatically demonstrated his unorthodox method of trying to get his messages across. He held a Bible up for us to see and began to rip out multiple pages at the time, utterly destroying the Bible. This was a tremendously conflicting image for me. It was frightening, but also somewhat liberating. It further reinforced my growing disdain for organized religion. This unorthodox demonstration accomplished two things in me; it emboldened my "Doubting Thomas" leanings, but at the same time it solidly infused in me an unquenchable fire to find the truth. I knew I needed to be totally at peace with my conclusion. It had to be ironclad. I would therefore be driven until I felt fully satisfied with the results of my search. Even with my intense desire I did not—could not yet—fully appreciate what knowing the answer would mean to me or to anyone who seeks and finds it.

For the remainder of my college and medical school years, if someone brought up religion, I would ascertain their spiritual belief, assume a 180 degree opposite stance, and draw them into a debate. This turned out to be a good strategy. I learned a great deal, but it would ultimately take much more to get to THE answer.

THE BEGINNING

Imagine with me a hundred trillion light years ago, give or take any other random number we choose as the beginning. This simply states that any point can serve as a reference time to the beginning of the cosmos while we search for our God. What can we theoretically, realistically, logically deduce? Before there was not even one simple pebble, what was there? Nothing visible! Albert Einstein gave us the clue. It was energy, still in its native wave form. $E=Mc^2$. From raw energy came everything there is and ever will be. Logic dictates this. Everything that is not energy is mass or matter, both of which are simply energy as non-solid waves that have been converted to something having mass/matter. This encompasses everything from a simple electron to a massive star (sun).

The idea is implausible that "poof the magic 'dragon'" (pebble, rock, lead, copper, silver, gold, platinum, or God) should simply appear from nothing.

As a scientist and a magician (Order of Merlin), I know, but in actual reality so does every intelligent person reading this, it takes great preparation, experimentation and practice to make even a silk scarf "appear." What about the rabbit from a hat, or ultimately... humans? For a magician, it takes a great deal of effort and practice.

Energy apparently is as ubiquitous as space itself. It is the most important and essential entity. Though ethereal, not even a substance or an element, it remains the one thing both native and essential to the entire cosmos. Without energy we have nothing. If a God wills something, it happens by the use of available energy. There is no other possibility.

A brief detour may be instructive. Suppose we look at Einstein's theory backwards, $Mc^2 = E$. Can you guess where I am going? Reverse engineering. This formula gave us the atom bomb from a relatively small mass dropped from the Enola Gay on August 6, 1945 which converted instantly to energy; enough energy to destroy a city with a population of 350,000, instantly killing 146,000, maiming and injuring countless others. A second bomb dropped days afterward, which essentially ended World War II. Sadly, the death count rose, as many others would later be counted as delayed casualties.

The foregoing is a dramatic graphic illustration of the original formula $E=Mc^2$ and how much mass we can develop from all the abundant energy floating in the universe.

Energy then is the foundational basis for everything of mass—and especially life. A body without energy is dead, whether it be plant, animal or human. The energy can be bound tightly, such as in a rock. It can be beamed tightly, such as in a laser, or loosely, such as in a flashlight.

While energy continues to be studied seriously by academic scientists, most of us in the world in general are mostly familiar with the kinds of energy from the sun and other traditional sources that move cars, trains, planes, rockets and the like, ... even ourselves.

As scientists we know a great deal about energy and have even discovered a different energy named dark energy. It is in space and is separate from the typical energy of which our five senses are aware. We don't yet know the extent of what purpose it has, but we know it has an important role in the universe. We know it is instrumental in pushing celestial objects outward with an awesome powerful force throughout the galaxy and the universe, but the full purpose of this has not yet been discovered.

We know that traditional energy in its original/natural state is a wave rather than a solid. Curiously, it has been proven that the wave form of energy responds to human consciousness, will or volition—changing to a more solid object such as an electron which has mass—as well as vice versa, from an electron back to a wave. This discovery is of huge significance. It's a game changer in understanding the difference between non-interaction with wavelengths without mass to mass without wavelengths simply by human consciousness. It gives new impetus to consider the age-old conundrum whether or not the mind can control matter. Particularly in the field of medicine, one would often hear the phrase "mind over matter" or "mind over body."

Of course, we now have proof; the answer is yes, at least to some degree. That will be discussed later. It should be noted here, however, that it gives new meaning to the power of the spoken word, the conscious word of will or volition. Not just as we think about it for ourselves but in what we have been taught about God. This explains how, if God speaks or wills it, it happens. The Bible tells us that the "Word was God." John 1:1 says "In the beginning was the word, and the word was with God, and the Word was God." Without volition nothing would have come forth—no thought, no order, no control, no laws of physics, no gravity, no life. Believe me, believe logic: nothing!

In short, simply put, the genesis of all order and control was the will of God before all gods, so many eons ago we may not even know how to count that high.

We have no plausible alternative explanation except that this simple discovery was the first discovery of the cosmos and made by the most basic assemblage of energy discovering itself. Certainly, the question arises, how could this happen?

No one knows the exact emergence of God. We envision God as Self-created. If He were created by another being or force, that would be a God before God, which is not Biblical or theoretically reasonable.

The critics will obviously argue for the impossibility but many of those critics will be the ones who claim that creation on earth is a happenstance. They claim a few molecules happened to be at the same place at the right time and formed an amoeba and from that amoeba has come all life. Consider how young our earth is (4.1543 billion years old) compared to other known entities of the universe (our Milky Way Galaxy is 13.21 billion years old) and going beyond to the birth of the universe (13.82 billion years ago). Consider the realm on the beginning of the cosmos at large, an infinity of time explained elsewhere. Our universe could not have come into being without the necessary mass of material needed for the formation of the universe, mostly regarded as by way of the "big bang." Energy did not suddenly make itself into a variation of different masses that could cause the explosiveness that rendered our universe so diverse in materials, and in so orderly a manner, awesome and beautiful to behold and contemplate.

We cannot make the mistake, as some do, and assume that time, space and everything began at the birth of the universe. Witness the foregoing paragraph. Ipso facto materials were necessary for the "big bang" or by whatever means the universe was formed and developed. That fact alone demands preceding cosmological activity.

Everything prior to the birth of the universe is completely unknown by us. We are projected into conjecture and theory based on deduction and extrapolation of what we have observed, studied and learned in this universe which is just a small niche in the greater cosmic expansion of space discussed elsewhere. But somewhere, and somehow, in that infinity of space and time the genius of vision, imagination, materials and grand design developed and was put into action. I believe that that was the first cause to our known universe. We have named it God. More on the conjecture of these times mentioned earlier in this chapter, will be discussed in more detail elsewhere.

If it were impossible over countless trillions of light years for this to happen, how can we claim to have developed ourselves in such a short time, and further, to develop electromagnetic brains that can perform rapid and reliable calculations—amazing brains that can be taught to learn as we have learned? We must concede there is a mind infinitely older and more knowledgeable than any of us, and that it has mastered the power of the word, of intention, of will and volition to its own advantage to improve upon itself through the eons.

If we are not blind and deaf to the various communications we have with this super intelligent being we call the supreme being, God, we should ever be grateful that at this point in history we apparently are one of his finest creations, no matter the route it took for Him to get us here.

The conundrum that we must be able to escape is to get out of the anthropomorphic mindset—that of trying to imagine God to be "like man." To be able to understand God and creation is in a sense to find a close parallel development of the Creator having come up in similar stages as we have, from humble infant beginnings. God replicated his developmental stages in humans but in flesh and blood, for a finite period, whereas He is on a cosmic scale and of infinite duration. Even so, we may also be looking at immortality preserved through our extracorporeal mind or soul.

The door is opened to understanding how even so many trillions of light years before us God had time to evolve to the point of designing and creating our universe with galaxies and solar systems. It is essential to know this, but ridiculous to speculate on a time when "God discovered Himself."

At this point it must be noted, God may have experimented with other galaxies or universes and they may be existent in distant parts of the cosmic space or they may have been destroyed. Our universe may be the only existing ongoing universe in the entire cosmos, but it is possible that others exist as well.

What is important for us to realize is that whether we are an initial universe or a present-day iteration of universes, we are apparently the high point of God's creation. After all, we are here in the present time. If that is true—and how can it not be?—we should be in awe of our position and make every effort to be in harmony with God and what our best perception of God can be.

As I write this, I'm sitting comfortably in a condo on the Gulf, overlooking the beautiful white sand beach heavily lined with umbrellas and populated with seemingly happy people. I see the vast Atlantic ocean out to the horizon, the moon slowly drifting by overhead. I also see the marvelous array of architecturally pleasing high rise condos and an occasional airplane flying overhead. How can I not be truly in awe of how much 'creature comfort' we have invented and made available to ourselves? These also are amazing, even miraculous accomplishments we have gained in such a short time here on earth.

Even more daunting is the knowledge that we are on an average planet circling one of 200 billion suns in our Milky Way Galaxy and further extended as a part of at least 200 additional billion galaxies in the universe.

"WOW!" cannot be written large enough.

How richly we have been blessed with reasonable health and longevity, intelligence, creativity, adventure, curiosity, doubt versus trust, discovery, desires, dreams, inventiveness, caution, love, empathy, sympathy, caring, and morality. More importantly, an innate belief in a higher power, along with an unquenchable desire to find and identify therewith, as well as free will and helpfulness. Unfortunately, along with the good is the bad, greed, dishonesty, violence, depravity, deceitfulness, immorality, temptation, rationalization, addiction, agnosticism and atheism.

Our search is hampered by several naturally occurring beliefs that arise and get in the way of reason and analysis. One example is the normalcy bias. Example, the sun comes up every day. We consider it normal and we expect it. We might see that most of our friends drink colas or alcohol so we develop a normalcy bias that it is okay. Our economy is moving along at an acceptable pace (at the time of this writing.) We consider it a normal thing and that is a bias and prevents us from seeing the signs of an approaching economic collapse.

We were endowed with discernment, but some seem to be able to use it better than others. Was that fair? Yet, total and correct discernment might have been extremely boring, or perhaps maybe we still need a little tweaking.

WHAT IS NORMAL? THE NORMALCY BIAS

One huge stumbling block for humans is our "normalcy bias." Our brains set as reality whatever we have known as fact, and experienced or practiced in our own lives as truth. The brain chemistry regulates this, especially the neurotransmitter Glutamate, which is ubiquitous in our brain. It is the arbiter of what is good or bad for us, pleasing or displeasing, healthy or unhealthy, spiritual or evil. Everything that is presented to us, the glutamate will judge to be either neutral, positive or negative. This is not quite the same as conscience, but the two may be in partnership in some dilemmas.

This normalcy bias is so strong in the human consciousness that its acceptance—and obedience to it—are almost as natural to us as how we see and feel about our skeletal foundation. We completely accept it and rely upon it without question.

Examples include whether or not we have a religious belief. If we do, which one, or which branch of it, and furthermore to what extent do we adhere to its tenets? Another great example is politics.

Which country were we raised in? Which political persuasion, through upbringing, becomes comfortable in that nurtured bias? Are we sports enthusiasts or not? Which sport, and which team do we most align with? What is our attitude toward health? The examples go on and on.

The teen years are critical. Some teens will "double down" and become essentially entrenched. Others will see the spectrum and investigate other options, wanting to make an independent choice. When their choice is different from the path of one's bias, it is going against a part of the self, causing anxiety as the result. That person goes through an experimental phase of testing the outcomes of his or her choices. This process will assist them to learn pros and cons of various paradigms—including their own.

These are the adventurers who believe in themselves but are grounded in reality. They realize doing the same thing over and over will produce the same results. They understand that for different results they need different actions, options or paradigm changes. They have discovered logic to be more helpful in their life. It overcomes false narratives and discovers truth.

These, however, will sometimes be chastised by their peers or even ostracized. Often, however, they become admired, even leaders to others—showing them a better way and a better life. It takes courage to challenge our biases.

The complacency of bias robs one of their aspirations, their sense of adventure, and their potential contribution to self and society. Often it pits us against our friends and neighbors.

Be yourself, be politely bold and see what you can accomplish!

Unfortunately, the glutamate can be hijacked by excessively high levels of dopamine, or modified to a change of opinion (judgment) by long term heavy exposure to an alternative "reality" (example, Stockholm Syndrome), or by a traumatic or very overwhelming experience strong enough to cause an instant reset.

*Examples:

1. Post Traumatic Stress Disorder or miracle.

2. Saul's instant conversion to Christianity on the road to Damascus. He became Paul.

3. C.S. Lewis' unique moment in life of Agnosticism that he became aware of the fact of his belief in God.

A bias can be like a rut. It takes effort to get out of it. Some people stay in their rut of "normalcy" because of parental pressure, or peer pressure. Others simply lack the confidence to explore new potential ways of being and living their lives. This is sad. Explore options with logic. Choose wisely.

PERSPECTIVES

Most who believe in God conceive Him to be the Creator, ruler and sustainer of the universe. For those readers who have not given thought to the vastness of His domain, an introductory perspective is offered here.

Our local community is dwarfed by our state, the state is similarly dwarfed by our nation, our nation by the earth, the earth by the solar system, the solar system by the galaxy… that by multiple galaxies of the universe, and so on, (up to potentially even possible multiverses) but eventually the cosmos at large—which contains it all. When we have stretched our imagination to its maximum, we still have only just begun to understand the first step into infinity.

Likewise, time is infinite and incomprehensible. Scientists have measured the age of the earth to be 4.1543 billion years. The age of our sun is 4.6 billion years, and its expected remaining life is another 4 billion years. The universe is believed to be 13.82 billion years old. But we are incapable of measuring infinity because it has been building for infinity, and continues its march from past infinity to future infinity, second by second. Consider that one trillion minutes equals 1,930,000 years. That's roughly 2 million years. These concepts are truly beyond our ability to comprehend.

One light year is equivalent to approximately slightly less than 3 trillion miles. Our sun is 93 million miles from earth—only 7 minutes away at the speed of light. Our moon is roughly 250,000 miles from earth. Light from the moon reaches earth in 1.344 seconds. Our closest neighbor star, that we can see, is 10 light years from us, or 30 trillion miles.

It took roughly 6 days for the astronauts to travel to the moon.

Astronauts traveling at the same speed the distance to our sun, would take 2,232 days or 6.12 years.

It takes 30,000 years for the same light that reaches us in 7 minutes to reach the center of our galaxy, the Milky Way.

The stars we see have their own solar mass with or without a planetary system.

Our galaxy alone has a minimum of 200 billion stars with some estimates going as high as 400 billion.

When we draw a circle, basically anywhere in the sky, using the Hubble telescope, we can count 200 billion stars within that circle the size of a full moon. Inside that circle would be 100,000 galaxies which means there are approximately 200 billion galaxies known to us, visible to us in the sky.

When the sun was formed there were other stars called sisters and brothers. Some of these are now within 300 light years of us, and there are around 400 other stars within 3,000 light years of us. Stars continue to be formed. It is estimated that perhaps half and up to two-thirds of all planets that will ever exist have yet to be born.

Donna Goldsmith, astronomer, writes in Scientific American, volume 23, no. 3, entitled "The Far Far Future of Stars" that "astronomers see no sign that life has affected the cosmos on a grand scale. That will change in the future. All the universe will become our garden." She quotes Freeman Dyson of the Institute for Advanced Study in Princeton, New Jersey, who reviewed this situation. The grand leader in this sort of cosmic speculation, he stated, "I think I have shown that there are good scientific reasons for taking seriously the possibility that life and intelligence can succeed in molding this universe of ours to their own purposes." He wrote that more than 30 years prior to Donna Goldsmith's article. That was at the time a grandiose claim by Dyson.

CONCEPTUALIZING SPACE – HERE & BEYOND

Understanding the unimaginable:

The chart below uses dollars as something tangible. It will dramatically make the point to move our thoughts to space. If we want to understand the power and scope of God, we need to discuss His territories, at least the entire scope of our universe.

	Week 1	Week 2	Week 3	Week 4
Monday	1	128	16384	2,097,152
Tuesday	2	256	32768	4,194,304
Wednesday	4	512	65,536	8,388,608
Thursday	8	1024	131,072	16,777,216
Friday	16	2048	262,144	33,554,432
Saturday	32	4096	524,288	67,108,864
Sunday	64	8192	1,048,576	134,217,728 (At 28 days)

Look what only 3 more days does to the total!

Day 29	268,435,456
Day 30	536,870,912
Day 31	1,073,741,824

Conceptualizing Power of Doubling:

Consider the above schedule in the context of one's salary—doubling daily for a month. In the shortest month of the year (28 days), the salary per day would increase from $1 to an incredible $134,217,778.

Even more astounding, with only 3 more days, (in a 31 day month), the daily salary would have increased to $1,073,741,824.

Now that we have a powerful way of rapidly conceptualizing the unimaginable expansion potential, we will use the foregoing model to go further than we have ever gone before in space. The universe as we know it currently extends further even than our general imagination. The farthest star that we can see with the naked eye is still in our galaxy and is approximately twenty-five hundred light years away, plus or minus five hundred light years. That is a baby step compared to the farthest star ever detected. Its name is "MACS-J1149+223 Lensed Star 1". This star is nine billion light years distant from us. More accurately stated, the light from that star has been traveling to us over a period of nine billion light years. That means that the star is actually currently much farther away from us.

It was seen with the Hubble telescope. Ordinarily even the Hubble telescope would not have been able to see it, and did so only by a unique happenstance. The light coming toward us had become too feeble to see but there was a galactic explosion, a supernova, between it and the earth. The light reached the galactic explosion which created a magnetic field which acted as a lens, diverting the light rays around it and converging them on the other side. This brought the beams closer together, thus magnifying the intensity of the original light by 2000 times. This was for only a brief period, and by an incredible coincidence, the Hubble was in just the right place at the right time to capture the image.

With this perspective, and a reasonable understanding of how vast and far reaching our universe is, add to that the doubling effect discussed above. Since space is infinite, we can do that. Let's double the size of our universe. Awesome, right? Now let's double that again, and then let's set up a program where every day we double the size of the previous collection of universes. Looking back at our monetary chart with one month of doubling we get the picture of how vast doubling the universe every day would be. But it doesn't stop at the end of the month, or a year, or a century. You can double for eternity and you will not run out of space because there is no option except that space is infinite.

NON-FATIGABLE TIME

Time marches on in a linear fashion—if we are speaking of what most people think of—known through the ages as Father Time. However, we now know that time is actually subject to manipulation. Using various techniques of traveling around the globe certain scientific measurements can produce different timed results of supposed equal linear time. I have read accounts of this, but I have not made an intense study of it. Carbon dating gives results in linear time, as do atomic clocks. According to our time apparatuses, we can accurately measure the age of the universe to be 13.799 billion years. The age of our solar system is 4.6 billion years old, and the age of the earth is 4.54 billion years.

What then could we estimate as the age of the entire cosmos to be, when it was empty of everything except energy? We have no way to understand because it has always been there. There was never a time that was a beginning. This means time also is infinite. So, we cannot go back to the beginning of time as an actual point. We can only project back to a time when space was perhaps totally empty.

How and why energy entered the scene I presume is a mystery.

Even as time passed through those early days, and through prehistoric days, it is now passing through our current days and it will continue to pass through the days of the future.

We can only understand time is basic, but it also has variable unique applications in certain circumstances. Hopefully, a better elucidation of this will be provided by scientists as more is learned.

The one thing we know is that time is infinite, retroactive and forward as well. Some have referred to it as the "eternal now."

SPACE IN GENERAL

It is not adequate to think of the universe as all encompassing. Scientists, astronomers, and the educated public often think of multi-verses. Literally, we could even think of quantum-verses, not to say there are, but there is enough space for them.

How can we be sure? Especially when some even speculate that space itself is minute, empty, invisible cubicles stuck together. It is difficult to believe such speculation could even be postulated as theory.

Space has to be equated to infinity. Consider that someone claims to have found the end or edge of space. What do we expect to be there, a wall, or an edge as if a cliff? If there is a wall at the edge of space would there be a door we could open to see what is on the other side? No surprise here. The obvious answer is more space. If you reach an edge and step over it, you will fall into something. What could that be, other than more space?

Accept reality. There is no other option.

Think of it this way. Earth is floating in the space of our solar system. Our solar system is inside of our Galaxy. We have one star, the sun. Our Galaxy has 200 billion stars. It is named the Milky Way Galaxy. Our galaxy is floating in the universe which is theoretically and objectively calculated to contain 200 billion galaxies (in which our galaxy is one.) Our universe floats in the great cosmos. What is to say there aren't multi verses of up to 200 billion universes and if that is the case, what is on the other side of that, except more space? You absolutely cannot expand your imagination further than the obvious that cosmic space itself truly defines infinity.

SURGEON THANKS JESUS

A prominent surgeon, strong in his faith, was part of a holiday gathering where a group of three physicians, including myself, and a couple of non-physicians were having a conversation regarding medical things.

In the course of the conversation, I asked the surgeon if he had ever been supernaturally "given" a diagnosis for a patient. At first he replied "no," but five minutes later he turned and said, "yes I did."

He explained, one day at work he picked up the chart and entered the room to see a healthy person sitting there. Her chief complaint was a scratched finger. He was highly specialized and was quite baffled that she was on his examination table. He even asked her, "how did you get in here?" Her answer was that she and her husband were traveling from the west coast area through Birmingham to Florida. Not far out of Birmingham she happened to scratch her finger on something metallic. She was concerned she might need to have it examined and possibly need a tetanus shot. They decided to stop at the first medical building they came to, which they did. The lady at the reception desk had assigned her to him.

He commented, "Well okay." He took care of her finger which was a mild scratch, nothing important. In the process he noticed she had a problem with the elbow of the same arm. He mentioned it.

She commented that it was nothing.

He said it was interesting, and he would like to know more about it.

She told him how the injury happened, what had been done for it, and that it was not a significant problem.

The surgeon remembers lifting his head and silently saying, "Thank you Jesus."

The background: One of his doctor friends had the same injury to one of his elbows but neither of them had been able to figure out what had happened. He talked to his colleagues, yet none of them could figure out the answer. This random patient, in relaying her own story, had been able to give him the answer they all needed. The interesting thing about this miracle is that it portrays what believers often take for granted and don't think much about. He remembers saying 'thank you Jesus,' but he had not contemplated the orchestration of events that had taken place (which had to have come through divine intervention) in order to have her make this trip through Birmingham, have the accident in proximity of his office, and the very unusual assignment of the patient with this minor situation to this highly specialized surgeon. The only explanation is that there was a supernatural design in getting the answer he needed delivered to him through the patient.

It points out that we need to be aware of all of the contacts we meet, and the interactions we have, for indeed they are quite possibly divine interventions and blessings we are being given, if we are truly paying attention. This story illustrates a seeming coincidence that would defy any odds, except that with God's intervention the fix "was in."

BRAIN / MIND

Does our mind and brain have any similarity to the brain and mind of Jesus? In an extremely limited fashion, yes. After all, he came to be one of us and to demonstrate to us the perfection of human living as well as coming for our salvation.

We must be thankful for the knowledge and the basic tools we have, and be humble enough to accept our limitations, whether they be more, or less than others. What Jesus did was not recorded solely in the Bible, but was also recorded and researched thoroughly in the secular writings of the time. The parable of the woman at Jacob's well, when thoroughly considered, is an awesome demonstration of the brain and mind of Jesus, and his ability to integrate the power of the mind with the human skills of the brain. As a result he was able to relate a powerful message to the woman and the disciples.

What about our brain? Does our brain or mind ever perceive things knowledgeably that are in distant places, things impossible to know that are in places apart from our traditional five senses? Yes! This actually happens quite often. In fact, in historical Germany this happened so often that they coined a word for it:

doppelgänger, meaning double goer, a person in one place knowing something that is happening in a distant place. This was prior to telegraph and radio.

We have read and heard of stories where a person would be presumed dead on the operating table only to be revived. Later, once they are awake and lucid, the person sometimes explains in exquisite detail what transpired during the time they were presumed dead, from both a visual and auditory perspective. In almost all of these incidents the events relayed were observed from a position of being above the scene, looking down upon it.

Another interesting example is that Leonardo Da Vinci painted an aerial view of the city in which he lived, when in reality he had no natural way to have been suspended to view the scene with his eyes. It must have been the Doppelgänger effect.

We now refer to this as the "nonlocal consciousness."

Personally, I have had one such experience that I will share, and although on a smaller scale it was profound to me at the time. Over time my experiences became more dramatic throughout my search for scientific evidence of spiritual reality.

I share my story as it relates to the foregoing parable to help describe such incidents, yet they remain unexplained.

Around the year 1970 we were hosting a neighborhood party at our home in Evans, Georgia. A neighbor had invited a friend from

Macon, Georgia to come with her. It was a typical gathering of young adults.

I passed by three women in conversation—my neighbor, her friend from Macon and one other lady. I caught the drift that they were talking about extrasensory perception or things of that nature. The visitor from Macon essentially derided any comments on that phenomenon being possible. Hearing this I simply commented that I was not sure, but that it seemed to me that sometimes things point to such happenings.

She appeared indignant, pulled her head straight up, looked me directly in the eye and challenged me. "Okay," she said, "you don't know me. If there is anything to that, then tell me about me." I was shocked at the challenge but without being aware of how I knew these things, I began telling her of her being married and the number of children she had... I don't remember all of what I said, but in a reasonably short order she almost yelled for me to stop. Looking quite troubled she commented, "I don't know how you know these things but please stop." This phenomenon was a unique occurrence for me, for it has not yet reoccurred. However, it was a profound revelation for me that such things could happen to ordinary people. I presume it was profound to her as well, and possibly opened her mind to learn more.

In my psychiatric work, often people tell me things that I might think are irrelevant to the issue at hand. I would silently wish that they would tell me about a certain part of their problem or life, so that I could gain insight into how I might help them. Not infrequently they would interrupt themselves to tell me exactly what I needed, and then would return to their previous topic.

The first story shared seems to imply that I was able to tap into and retrieve cosmic information, while in the case of my patients answering my silent questions, it could have been a form of mental telepathy, person to person. In actuality, we don't know which it is, or perhaps if they are not one and the same, after all. Whichever the case, pursuing the answer is intriguing and I believe would prove very interesting.

I'll share another example. When my youngest child was approximately four years old, she, her mother and I were having breakfast. Everything was quiet. No one was saying anything. As a total surprise our daughter looked up from the plate and simply announced "the grass is high" and resumed her eating. My wife was visibly shocked. She had just that instant had the thought that she should call our yard person to cut the grass because it was getting high. This is as proof positive as we can get of mental telepathy. We just have not yet been able to find a machine to capture and test that reality.

It appears logical to deduce that the brain, or mind, has mental telepathy ability.

Equally evident, the mind is independent, and functions separately from the brain, but can and does exist in a manner compatible with the brain. The brain is the host to the mind. The mind is extracorporeal (consisting only of dark matter) and can exit the body and return. This is a huge genius introduction to our body that is little known. Being exclusively dark matter, it is totally untouchable by anything that we know of our universe. If a Hydrogen were to explode just 10 feet away, our body (and the brain) would be completely destroyed but our mind would be unaltered and left completely intact.

Therefore, it follows that no matter how we exit this world, our body will be discarded, but our spirit Mind will have our totality dutifully recorded. We will be conscious of our past, the best and the worst.

Think about that.

A DIFFERENT LOOK – A PARABLE

While Jesus was in Samaria he sat resting by Jacob's well at about the 6th hour when a woman from Samaria came to draw water. Jesus said to her, "give me a drink," for his disciples had gone away into the city to buy food. The Samaritan woman said to him, "How is that you, a Jew, asked for a drink from me, a woman of Samaria?" for Jews have no dealing with Samaritans.

Jesus answered her, "If you knew the gift of God, and who it is that is saying to you, 'give me a drink', you would have asked him, and he would have given you living water." The woman said to him, "Sir, you have nothing to draw water with, and the well is deep. Where do you get that living water? Are you greater than our father Jacob? He gave us the well and drank from it himself, as did his sons and his livestock." Jesus said to her, "Everyone who drinks of this water will be thirsty again, but whoever should have a drink of the water that I will give him will never be thirsty again. The water that I will give him will become in him a spring of water welling up to eternal life." The woman said to him, "Sir, give me this water, so that I will not be thirsty or have to come here to draw water."

Jesus said to her, "Go, call your husband, and come here." The woman answered him, "I have no husband." Jesus said to her, "You are right in saying, 'I have no husband,' for you have had five husbands, and the one you now have is not your husband. What you have said is true." The woman said to him, "Sir, I perceive that you are a prophet. Our fathers worshiped on this mountain, but you say that in Jerusalem is the place where people ought to worship."

Jesus said to her, "Woman, believe me, the hour is coming when neither on this mountain nor in Jerusalem will you worship the Father. You worship what you do not know; we worship what we know, for salvation is from the Jews. But the hour is coming, and is now here, and the true worshipers shall worship the Father in spirit and in truth, for the Father is seeking such people to worship him. God is spirit, and those who worship him must worship him in spirit and truth." The woman said to Him, "I know that Messiah is coming (he who is called Christ). When He comes, He will tell us all things."

Jesus said to her, "I who speak to you am He." Just then his disciples came back. They marveled that he was talking with a woman, but no one said, "What do you seek?" or, "Why are you talking with her?" So, the woman left her water jar and went away into the town and said to the people, "Come see a man who told me all that I ever did. Can this be the Christ?"[8]

This teaching plainly allows us to see Jesus' direct identity with God as One. He answered that which she did not ask and provided life details impossible to know otherwise, and not only that, He identified himself as the Messiah, the Christ speaking as if God Himself.

At the unnoticed ending to the parable, she emphasized that the reality of the flesh is to continue to need water for the flesh but she left the jar for Jesus and his disciples and went away without water and spread the living water to the townspeople.

What is the scientific message of the mind in this parable? Either His mind is able to communicate with her mind, certainly without her conscious awareness, and ferret out the information His mind desired. This is possible if the mind truly is extracorporeal although living, mostly with the corporal brain as its host.

The second possibility is that as His corporal brain was conversing with her, His extracorporeal mind was in direct contact with the spirit mind or knowledge bank of God, therefore, being as if it were connected to the archives of God's knowledge, or a recipient

[8] John Chapter 4

of God feeding him the information he needed. This brain/mind phenomena is discussed in further detail in the previous chapter.

ALABAMA MIRACLE

Catherine was telling me how much she liked the church she was attending. She particularly liked the music played by the band in which she was a participant. She both played the piano and the guitar. Her faith meant a great deal to her and it was important that she let me know that.

I asked if she had experienced any miracles in her life. She answered, "Several!"

I asked if there was one she would like to share. She said, "yes, and this is my favorite miracle," and shared the following story. She had always enjoyed playing the guitar, and since high school, she had been fascinated with them: the different looks, acoustic sounds, price range, reputation, and prestige. She found one that seemed perfect for her. It was beautiful but beyond her price range. When she strummed a few chords, she had to try a few tunes. Never had she heard such heavenly music from a guitar. She knew she had to have it as her own.

Catherine considered her budget, and every item was parsed as much as possible. She saved as much as she could. She found ways to make extra income. After many months she was able to purchase her favorite instrument. It gave her great joy to play it, and a sense of pride that she had not only found the perfect one for her, but had personally found a way to work and pay for it.

Some time later, (and a few years back from relating the story to me), a tornado came through and hit their house. She explained that her sister lived next door to her, and they both had been listening to the weather report because tornadoes were in the forecast. One was headed their way, so they took shelter in their shared in-ground storm shelter.

After securing themselves inside the shelter Catherine remembered that she had left her favorite guitar unprotected in the house. She had not even put it in its case. Given its significance and sentimental value, she was especially distraught that she had left it in the path of danger and was very upset.

Hearing noise from the approaching tornado she became more frightened and started praying for the safety of her guitar. The louder the noise the more fervent and passionate her prayer. She was not worried personally for herself because she was in a safe shelter, but she knew the guitar was exposed in the house.

To their horror the tornado sucked the roof off of their protective shelter. They were drenched, but unharmed.

Simultaneously the roof of her house also was ripped apart. The house was considerably damaged. She continued to pray that her guitar would be safe, but when she saw the wreckage of the house she worried that it might have been damaged or destroyed altogether. She continued praying until she was able to get inside the house to find her guitar.

Despite her faith in prayer, she still was surprised when she located the guitar. It was exactly where she had left it. Leaning against one of the walls, which was still standing. A piece of the ceiling had fallen and landed in front of the guitar, leaning over it and against the wall, providing shelter. The guitar had not moved one inch and had no damage, not even one drop of water. The guitar case which she wished she had left it in, however, had been damaged, but fortunately was still usable.

Catherine has told this miracle to several people. Her faith was always strong but is even stronger because of this miracle.

MY DAD'S MIRACLE: THE PREACHER MAN

During my college years I elected a pre-med curriculum. My parents rebelled, saying in effect, that that was not acceptable. Expecting them to be joyous at my decision I was disappointed in their dissatisfaction and understandably wanted to know why. They asked me to remember them telling me that as a two-year-old I became extremely sick for a long time. They had been convinced that my life had been spared after they had prayed and pleaded for my life. I confessed that I did remember.

There was more to the story, which they were now obliged to reveal. "When we had prayed so long, and you did not seem to be responding, your mother and I decided to promise your life to the service of God. Our thinking was that you would grow up Christian and serve as a missionary."

I responded, "I am grateful but I don't want to be a missionary. Besides, I can serve the Lord as a doctor."

"It is not the same," Dad said, rather firmly.

I persisted with my reasoning, but he did not relent. Relatives pointed out that my goal was impossible to reach; no one on either side of my family had ever even attempted such. The effort would be foolish and a waste of time.

My parents became guilt-ridden feeling sure they had let God down. They decided they would have to pay the price, and make amends for their failure to deliver on their promise to God. With his third grade education, my father drove sixty miles to Mercer University, in Macon, Georgia and enrolled in seminary classes. He continued to persevere in this round trip twice weekly until he successfully completed enough courses to qualify to be an ordained Baptist minister. A regional official agreed, and he was ordained.

Prior to this he worked a series of jobs, but nothing seemed to last. Once he ran a service station and later a grocery store. He worked as an insurance salesman and during the depression years he worked on the WPA, but now he was a minister.

He founded a church for the less fortunate, grew it to the point he was able to build a church with Sunday school rooms, a kitchen and a good sized sanctuary. He pastored his "Faith Baptist Church" until strokes and dementia incapacitated him.

Fortune allowed me to visit his church, listen to some of his sermons and observe my father, a man of faith, supported by my mom, a woman of faith, who played the piano by ear for their congregation. The townspeople came to love and respect my father as a man of God and referred to him as pastor, reverend or to some simply "the preacher man."

The above was impressive and certainly emotionally satisfying to me, seeing my parents sacrifice so much in service to the God they had taught me to believe in. However at that time in my life I was struggling to reconcile what I was learning (mainly science based) with the faith I was brought up with, and I was beginning to have doubts.

At the very end, in his hospital bed, suffering from his final stroke he became comatose and finally there was nothing left of him but his struggling body for three days.

In my presence and other family members he suddenly opened his eyes, raised his head and to the top of his voice uttered the words, "Praise God."

Although he never knew the meaning of the phrase "double entendre," he ended his life using it to perfection, proclaiming his faith for me and everyone present to hear as a command. He was simultaneously meeting and greeting his creator, and exhorting all who would come to know his last words, to follow him in the praise of God. I must confess that it was a powerful experience for me, yet with my vigor, self-sufficiency and especially my fascination with science, I remained a doubter. Were God and faith a human invention for our comfort? Would science even be proving there was no such thing as a God?

Now, coming full circle, I know and believe that my drive to find the truth and make it understandable was not only a personal quest, but also an answer and an atonement for my parents' promise to God that for my life they would raise me to serve Him.

As I now replay the foregoing scene I marvel, once again, at God's many ways, often surprising, to speak to us, to say, "Hello. I am the I am. I am not dead. I will not, cannot die. I am eternal for all that came before and all that is to come." Harmony with God is also harmony with Christ and the Holy Spirit which is God's essence and presence in every breath we take. I thank Him for my parents and my many blessings.

Addendum: An animal analogy:

Since part of this book is to demonstrate God's imprint and influence throughout nature, I take this opportunity to remind myself, and you, of the many stories where animals have befriended humans and each other. We take heart when we see films of such happenings. We cannot help but notice human-like kindness between the animal world, and not limited to one species. We see an animal nurturing an infant orphaned from a species different from their own. In at least one case we know of a human infant that was nurtured by an animal.

With His last breath my father praised the one he loved and worshiped. I'll now relay an incident that demonstrated the equivalent in the animal world.

My wife and child desired to have a dog and wanted to start with a puppy. We found a female Sheltie who turned out to be a princess of a dog, so we named her Princess. She was totally devoted to the family, but especially to my wife and our daughter. It was a natural bond that was normal throughout her life.

At the end I came home from work one evening and she signaled to go outside. I knew she was sick and feeble, but she managed to negotiate the three steps both down and back up. Back inside, she went to the sitting room and lay down. She stayed in that position. Eventually we realized she was apparently waiting there to die. We sat with her for maybe two or three hours. My wife sat directly in front of her and I to the side. She lay motionless all that time until eventually I saw her front paws start pushing herself upward. She managed to lift her head high enough to reach my wife and kiss her on the lips saying "goodbye, thank you for a good life." She drew not even one more breath. She died before landing motionless on the floor. This is a dog doing the same thing for my wife that my father did for his heavenly father, by finding a way to express "you've been good to me. I thank you. I appreciate you. You made my life joyful and now I say goodbye."

DOUBTER OR BELIEVER

This book is about God, proof of His existence, divine intervention, and a rational scientific basis for a God.

Those who believe by faith will likely disclaim any need for a treatise such as this. My close friend, Samuel A. Saxon, Ph.D., clinical psychologist said, "Proof—who needs proof when you have faith?"

The Apostle Thomas (known as 'doubting Thomas') needed proof. Fortunately for him Christ reappeared with the proof. I—and many others like me—have for various reasons, been more like Thomas. Agnostics are more analytic and scientific, and basically live with an open mind.

Atheists are the antithesis of the faithful. Theirs is an opposite conviction; they believe there can be no God. The ultimate question for an atheist is this: IF scientifically and mathematically it is proven that God (a creator) exists, then which of the following would they choose:

A) "I reject the scientific law that math proves everything."

B) "I am looking at proof that I cannot believe, so I will be damned."

C) "God's existence is proved. It is rational and I therefore join the kinship of believers."

I confess my unwavering belief in a divine, interactive, creative Deity. I did not quickly or easily arrive at this conclusion. In this book I share bits of experiments and knowledge learned that cast doubt on my belief, and I also share equally strong experiences and occurrences that I challenge anyone to explain in any other way except for Divine intervention. In the end the latter experiences won out, leaving me with great peace.

Believing should have been easy. Daily my believing parents lovingly preached belief and faith to me. This was at a time when there was no such thing as TV and for quite some time we even had no radio. Not everyone could afford one. Most teaching and learning was from the parents, the church and school, so naturally, most children followed obediently in the paths set forth for them by example. However it wasn't that straightforward for me. With a naturally inquisitive mind, hours and hours of independent reading, going to college and learning the scientific method, simply conforming to my parent's example was not satisfactory to me.

Organized religion besmirched the spirit, proclaimed by religion. Agnosticism seemed to be the only position during those years.

But because it was anathema to my upbringing it created a state of spiritual unrest. I worked at it through debating with anyone who would discuss religion, myself taking whatever position was opposite theirs. Although I had learned a lot, it seemed to me that it was mostly dogma with no solid answers. After marrying and having a child I at least knew children raised in a spiritual environment had a better chance of having good character—and this was usually associated with believers. I decided to raise my children in the church. To be hypocritical would be unacceptable.

So I asked God, "If You are truly there, please show me."

He did.

SETTING EXPECTATIONS

For this book to serve its full purpose, readers at the end should accept:

1. Divine creation must be unshakably believed, and hopefully completely irrefutable.

2. Scientists must accept with humility that what was proof positive in the past is sometimes undone and supplanted with new, even more proof positive discoveries. How can we have complete confidence until we fully comprehend one hundred percent of the cosmic energy/matter equation. Perhaps we are only now beginning to possibly recognize and openly admit collective glimpses into and theorize about the 95.1% unknown—called dark energy/matter.

3. We cannot deny what we perceive and know, even though we cannot yet capture these things with material instruments, no more than the lower animals can tell themselves or discuss among themselves what is the composition of wind. They only feel it and respond.

4. Believers by faith should find added comfort and strength. Additionally, they should strive to gain information to share with doubters.

5. For doubters, a strong compelling rationale is to give God a try. Seek Him. You should be glad you did.

Humans, some more or less than others, feel and know experiences from outside our constricted 4.9%. Examples are ubiquitous. A few examples, both simple and complex, will be presented.

Examples I will share as you read on;

In a third world community in crisis, a miracle happens.

An American minister's car breaks down, a miracle happens.

And many others…

A HUMAN CONUNDRUM

My parents and church raised me with hard-shell beliefs.

1. The Sunday faithful professed these values and beliefs for all to see. Yet my weekday observations of these same people were not convincing. In fact it had the effect of disillusionment as I began to notice that the "Sunday faithful" were not living lives that demonstrated the very values they professed to believe. This especially combined with scandals of some high-profile evangelists, turned me to agnostic thinking. Therefore, I began to fervently debate religion—pro or con, "real or not real?"

2. I came across the book "The Great Pyramid: Man's Monument to Man" by Tom Valentine. The great pyramid built six thousand five hundred years ago, wrote unmistakably in stone the exact date for the birth of Jesus Christ. The exact day of His crucifixion. The king's chamber was built so that on Christ's birth date a skylight was provided pointing to the space where the Star of David would be at that moment. This was astounding to me.

At the time I read this, the scientific model basically was ruling my life. The world around me was touting agnosticism, (even pushing atheism such as the book "God is Dead" by Ron Currie.) For me, atheism was unappealing, even anathema. Later I understood this was due to the "normalcy bias" residual teachings during my formative years. So, agnosticism felt comfortable to me.

This revelation presented me with a true conundrum: how could a science-based prophesy with such uncanny accuracy occur without divine design.

This question has begged answers by many, but no one brings the answer. (Discussed later.)

When one is a scientist and seeks the truth, one is bound by it. Interestingly the answers change as scientists learn more. The earth is not flat or the center of the universe. The speed of light is not the fastest speed possible. Currently quantum is the fastest. When one is a person of faith one has no less fervor for the truth and likewise seeks it, but their proof is in the heart. This method likewise has its challenges, usually in times of personal difficulty, tragedy, misfortune etc. In either camp we might ask ourselves, "Am I spiritually schizoid, not comfortable or satisfied either with or without belief?"

Until now!

Collectively science has proven much of what the Scriptures tell us. The evidence of the source is overwhelming. Collections of experiences by individuals throughout history, including myself, are overwhelming in evidence of a force beyond human capability, and an intelligence behind that force must exist. In fact, logical analysis of the facts of our world and lives points not only to our creation, but to our ongoing care and nurturing by an intelligent loving divine presence.

It is said that whatever can be proven can be proven mathematically. I believe that statement is true in proving the existence of a creator and a continuing interactive force palpable with all living things, but understandable only by the human brain. Therefore, I hold out to the reader, examples that follow—both faith based and science based—throughout this book that will lead to compelling answers and theories.

One simple experiment poignantly upholds the Scripture "and the Word was God"[9]. In this scripture, "word" is expressed as a verb as it means the Will of God, and the Will of God is the volition which is the activator. Whether spoken out loud or not it is identified by the word when we say work, we are expressing our volition for work.

Science has demonstrated conclusively that the mindful desire for a scientific event to occur does cause the change desired by the scientist making that decision. An example would be the motion

---

[9] John 1:1

of an electron that was activated by the thought or volition of the observer.

MIND-BRAIN RELATIONSHIP

The mind is different from the brain, but the word mind is often used incorrectly, interchangeably as brain. They are not the same. If you didn't have the brain, could you have the mind? Yes, but it would be severely handicapped. The mind needs the brain to interpret and employ knowledge and action, provided from God's directives. The brain needs the mind for all except the animalistic functions.

The mind-brain function is genius. It alone should remove any doubt of an intelligent creator from any person who studies it.

If we did not have both a mind and a brain, we would not have so clear a communication with God. It should be no surprise that we have a mind separate from our brain. The mind is designed to function independently and separately. However, it works in a compatible and complementary way with the brain. I theorize that one of the mind's several functions is to act as a spiritual homing device, though not interfering with free will. It is an intervening tool used often by God. Notice the drama. The mind is often important in miracles. See the chapter entitled "The Gospel Singer's Miracle."

Most of us consider we are using our mind to think through problems and concerns. We want to understand that which is greater than ourselves. We ask, "how does this work?" We use our brain, even though our mind may or may not be involved in the thought process. It certainly helps us toward the spiritual path if we seek it.

The brain is a flesh and blood machine. It functions much like a metal and electrical machine. It calculates, stores memory, processes and produces results. It tells a body how to function, or when to run. A computerized machine might act similarly. The functionality is dramatically analogous, except that the electrical machine is much faster. Flesh and blood functions at high speed but varies with age. For example: processing a thought requires 67 milliseconds in our twenties—the fastest, 89 milliseconds in our forties, and 110 milliseconds in our sixties. There are numerous tables on Google.

We observe other phenomenal differences. The brain cannot leave the body and function. The mind functions equally well inside or outside of the corporeal body. We have recorded proof, many times over, of a person supposedly dying during a surgical procedure but resuscitated. Later when awake and lucid, they describe what happened around and to their lifeless body with astonishing and exquisite detail. There is no logical way this could happen unless the mind is separate from the body. Therefore, ipso facto, we have a brain that is corporeal, part of the body, but we also have an extracorporeal mind that is not material and not restricted to the brain or the body. It is demonstratively independent in space. The "extracorporeal" mind can travel and go places where the body does not follow. There are instances of this happening throughout history which have never been previously explained adequately, although they've been described in many different languages. For example, the German word for this is doppelgänger, which means double goer. Science now refers to this phenomenon as "nonlocal consciousness".

One experiment attempting to study the speed of brain function misinterpreted and claimed it was quantum. We have no direct way to visualize or record the mind which does function at quantum speed and therefore the brain has the information imparted to it in quantum speed.

We deduce logically, when the physical or corporeal body/brain dies, the extracorporeal mind is freed and remains intact.

When the doppelgänger mind travels it sees without eyes and hears without ears. When it returns to the corporeal brain there is instant incorporation of data downloaded from the mind. The corporeal brain is integrated so that it reproduces what was seen and heard by the extracorporeal mind and transcribes it in an understandable language. The download can be in-part or in whole at the discretion of the mind, which is always in tune with the Holy Spirit guaranteeing God's Will with the mind, while not interfering with free will.

Recall the opening miracle. It reveals a personal encounter which not only proves this point, but amazingly proves that extracorporeal minds can communicate in perfect detail. This can happen when one is from a currently living body and the other is from a long-deceased body.

The mind/brain is in two parts: one part (brain) is intracorporeal and the other, (mind) extracorporeal. The intracorporeal brain works with the flesh and blood, and the electrical circuitry of the physical brain and body. The extracorporeal mind is ethereal, has consciousness and all the senses of the brain and human body and more.

Many people have experiences that can only be explained with an extracorporeal mind. The additive effects build an overwhelming mountain of evidence for its existence. The "Multi-purposed Miracle" is one example. I invite your close attention to the story and its details. [10] Like nonlocal consciousness, the mind is functioning differently from the brain, but is not traveling; the cosmic consciousness or God's intervention with His 95.1% abundant energy spectrum can directly communicate with our mind, consequently our brain. God, Jesus, and the Holy Spirit can instantly communicate with the brain through the mind. Look for examples of this as you read this and other books about miracles.

---

[10] Why consciousness is not the brain. Issue #36 New Edge Science Larry Dossey, M.D.

The mind, as an interesting fact, is not a part of our universe. In fact, it is formed out of the dark energy. It is immune to any effect from any source in the universe, It cannot be destroyed, even by a hydrogen bomb. The separate energies cannot recognize each other. It is tethered to our brain and dutifully records everything of our life. It is our soul and how we face God intact, no-matter how we die or are destroyed. It is yet another demonstration that God has thought of everything.

NON-LOCAL CONSCIOUSNESS

William James, father of psychology in America, was one of the first modern thinkers to endorse an outside the brain view of consciousness. This was espoused in an 1898 Harvard University Ingersoll series lecture. He spoke against such notions as consciousness being an internal brain function, or the brain producing consciousness, as "Cerebralism".

He used an analogy to make his point. When the brain is dead, we cannot further see consciousness in a person. It doesn't mean it wasn't there. It only means the brain didn't originate it or control it.[11] He stated for example, if you smash a radio, although there is no other voice or music coming from it, it does not mean that the source of the voices or music was dead or lost.

[11] My elaboration.

A CHILD'S OBSERVATION

One very interesting example comes from a child who reported an afterlife experience. Such occurrences are sometimes perceived with doubt. After coming back to life, being a child again, when presented with a coloring book and crayons, after looking through it, she innocently asked, "Where are the other colors?" It was a purely natural question, based on her remembrance of the other world—the world on the other side of what we perceive as reality. The question would be considered strange if one was not aware of her travel to the world beyond.

It is easy to miss cues from the spirit world unless we have our eyes, ears and mind open. Normalcy bias plays a role here, but we know that our colors are relevant to the spectrum of our light rays. If we had the opportunity to see the colors of the rays in the 95.1% of energy missing from our earthly reality, certainly other colors exist. Likely in abundance.

Logically thinking this should not be surprising. Energy is wave form, not matter. If 95.1% of energy is blocked from the human five senses but presumably not from spiritual beings, the spirit who visits the other side and returns will have a recording in their mind ,the soul, of what was witnessed, and often apparently deposits to the host brain letting it become, to variable degrees, part of the human memory.

These glimpses show us the reality of the spirit world and the joys we may expect to find there.

These teasers are part of God's way of showing us "I am real" or perhaps simply "I Am."

DELIVERANCE

Finishing the manuscript of the book you are reading, I approached my good friend and colleague, Sam, to favor me by reading it for content and providing me an honest review. Sam is a Ph.D. psychologist who has a stellar career in treating children and adults. Everyone who knows him is aware of these things about him: he is witty, funny and a devout Christian. I knew he would be perfect for giving me a critical review. With his pleasant demeanor he readily agreed stating that he would enjoy reading it and giving a review. His agreement for the task was enough for me, but he went further. He told me of a miracle he witnessed many years ago.

It was then I realized that the book was incomplete, as it needed one more miracle. Upon hearing his story, I realized there was no comparable miracle in the book, and I knew it had to be included.

Quite a few years ago a pastor named Bill French founded Deliverance Ministries. He was an invited guest to a small group dinner with a select group of Christians. Besides himself Sam remembered there were three elder Presbyterian church members present, another clinical Ph.D. psychologist, a physician with his wife who walked with a limp due a birth defect. One leg was three or four inches shorter than the other.

During the evening pastor French noticed the physician's wife limping. He approached her and asked if she would like him to pray for healing of her leg. She answered yes.

The pastor had her sit with her legs on a chair. where he could hold her feet. He furthermore asked all that were present to touch each other, and all pray with him for her healing.

Sam, though skeptical, participated willingly. After a time of praying. he noticed that the short leg retracted shortening by another inch. Later it protruded back to the original length, but then incrementally it would grow approximately an inch at a time until it had passed the other leg in length. It retracted to the exact length of the other leg and remained in that position.

After the session, the lady walked without a limp.

Of course, this miracle was something Sam continues to marvel about. Now, so do I.

THEORETICALLY SPEAKING

I will now share further thoughts on non-local consciousness.

Throughout the book I have shown the reader miracles, both personal to me and also from others. In general, they demonstrate God showing us He is in touch with our needs.

Our mind is the equivalent of an advanced indestructible microchip but has no tangible mass. How do we know this? We know from countless individual reports where the mind temporarily separates from the body, yet the body is alive and is rejoined by the mind with the new information. We know that generally it resides in our body hosted by our brain and is thereby connected to our entire body. It is known to, at times, detach from us, and "travel" where we physically do not go. Sometimes short distances such as hovering about us in an operating room when the body has temporarily lost the energy and breath of life, but the surgical team is maintaining vitality through resuscitative measures. Sometimes the mind exits the immediate area—such as the operating room—and proceeds to where loved ones wait with intuitive dread, and shares a message of encouragement that can be conveyed by a thought or feeling of peace or relief.

You will see examples of this throughout the book.

In rare cases one can receive detailed accounts of the full interaction between the spirit (cosmic) mind's interaction with the visited person. Such was the case with Edgar Casey who had hundreds of visits outside his body while his body remained in a trance.

The mind therefore is not affected by gravity—or any other barrier to instant travel—or physical connection to its host brain.

That travel is consistent with quantum speed, just as scientists have observed in non-local, immediate communication between electrons.

The mind is much more than an electron/wave phenomenon. It is a living being, obviously a part of the 95.1% of the energy matter, and it is communicating information to the 4.9%.

It can and does touch us directly, at least by communication, but we cannot in any way materially touch it. We can only meditate on our thoughts in our brain. However, it, the mind, knows our every thought, so meditation is a good communication pathway to the mind.

I theorize that while God has given humans free will, God also gave us a mind to be an ever-present conscience, but not the authority to control our will.

We collectively work toward space entities that can travel to and from Mars and other celestial bodies bringing back or transmitting back what strange and perhaps wonderful things that could have been found. Theoretically our mind could already be doing that but likely that is not allowed as we have not one such report.

Presumably, such information is cosmically available. Did Leonardo Da Vinci and/or Nostradamus somehow break through that barrier? Could the vast knowledge and the contributions of the great composers, artists, and inventors of all time perhaps be a glimpse at God's infinite knowledge? Imagine what could be possible if more people learned to access this part of their minds.

Did my mind glimpse beyond and breach the barrier for a timely multipurpose message delivery? Of course, it did. What other explanation could there be except as I unfolded it in my analysis revealed to you.

The mind has many advanced attributes, some used regularly. Some are specially reserved for uncommon needs. God—through the mind—can and does at times force control of one or more of the channels. This is as real as our favorite program being interrupted by a session of buzzes and followed by, for example, a bad weather alert. One good example of this is seen in the "gospel singer's miracle" shared in this book.

DOCTOR / SCIENTIST RELATIONSHIP

A medical doctor who is not a scientist is more of a medical technician, equipped with a set of diagnoses and a list of treatments, who then simply matches these with the patient's symptoms like a puzzle. This insight into doctor-patient relationships will be helpful not only in reading this book, but also in relating to your doctor.

All medical doctors have basically the same training, but lately that contains less of the scientific method and basics. The medical doctor who has an understanding of the scientific method will use it as one part of the overall assessment, as an additional tool. Every patient must be seen with a conscious understanding that, not only do they have a unique fingerprint, but they are unique in every way: personality, genetic makeup, chemistry, nurturing, education received, and their life experiences.

Everyone responds to treatment individually, particularly chemical medicines. All medical doctors must keep their minds open, not just to every new patient but to every single visit of each patient. The training of medical doctors does underscore thoroughness and compassion. In most cases this is generally a product of an individual's basic personality.

When more than one company makes a generic medication, each company's formula has something slightly different from the others. This slight difference sometimes causes a significant difference in how the (supposed) same medicine would affect any given patient.

One example will make the point. Mr. Stevens, a 45-year-old Caucasian male came to me for yet another opinion. He had seen several doctors before me and had taken their medications without positive results. Furthermore, he received no positive results with any of the medicines I prescribed. He faithfully came every three months to see if any new medicines were available and to discuss his situation. This went on for several months. Eventually Prozac was introduced. Samples for a one-week supply were provided. I suggested that he try it by taking one per day. If he had no side effects from it, and particularly if he felt any improvement, he should call me at the end of the week and get a prescription. Three months went by before I saw or heard from him. I was disappointed and apologized to him. I was sorry it had not worked. Since he had not called, I had therefore presumed that that was the reason.

To my utter surprise, he said, "Quite the contrary, Doc. I took one of those pills when I got home. The very next morning I woke up feeling normal. I had absolutely no depression. I felt great. It lasted a whole week before I felt depressed again. Each one of those pills lasted me a week so I didn't need to call you." He stated that he was still feeling normal and had no depression.

From years of experience and constant observation of what happens with patients and medication, I could immediately see this was going to be a quite different medicine for some people. A world of information poured into my brain just from that one patient experience.

The opposite can and does also happen. A primary care doctor prescribed the medicine for one of his 35-year-old female patients. She felt better the next day. The day after that she still felt better but quite nervous. The following day she was extremely nervous, anxious, and hyper and called the doctor. He increased the medicine by doubling the dose. The following day after the double dose she was unmanageable, was talking amazingly fast, and was illogical. She was taken to the emergency room. I was called to admit her to the psychiatric unit.

During my examination of her, she told me that she was fine. She had never felt so good. She excitedly told me that she had triple indemnity insurance. She planned to run her car into a bridge abutment so she would be able to collect her triple indemnity payment. She would pay off all of her debts and be fine. She was looking forward to doing that. Logic made no sense to her at that time because she was in a manic-psychotic state, driven by the medication. After a few days she was back to normal and was grateful, but she could remember what she had been saying and doing and realized how psychotic it had been. While I tried to convince her that the medicine would be fine if she would just take one per week, her experience with it had frightened her so much that she declined to take it at all.

The foregoing is to share the nuances in patients, doctors, and medicines. It is important to be a science focused doctor, keeping mindful of these differences and keeping abreast of accumulating knowledge.

The above example addresses the physical-corporal part of the person. There is also another part of the human which is the extracorporeal-spiritual part of the person.

To treat the whole of the human we must be scientifically cognizant that our patient is a physical/spiritual mixture. When the physical body is healthy, functioning harmoniously in all its parts, and the Spiritual entity is functioning in its Spiritual harmony, only then is the human truly in good or proper health.

A doctor grounded in both disciplines will generally have an edge in arriving at a complete diagnosis and appropriate treatment options. The medical and physical aspects are generally where we start with our patients, but too often this is the only part of the person addressed by their doctor.

Approaching the spiritual life of a person can be challenging, even if done gently and mindfully. It may be perceived as awkward or as challenging to some patients. But often it is received quite enthusiastically.

In many cases, the doctor does not know if the person is a believer, a doubter, an unbeliever, an agnostic, or an atheist. Not knowing this and not being able to comfortably elicit that information may be a serious and confounding factor in arriving at the appropriate diagnosis or prescribing the correct treatment plan. I have included a couple of harmless questions on the topic in my introductory questionnaire, which I find to be helpful.

It is noteworthy that a growing number of physicians are writing about their relationship with God and Spirituality in general. It appears to me that the physician scientist is educated scientifically therefore trained to recognize reality. The physician works constantly with the miracle of life, sees benefits to human health derived both from spiritual nurturing and scientific benefits. This compels us by desire, conscience, and duty from medical and spiritual directives to share what we have gleaned on our privileged path, to all who desire to know what we have learned.

SPIRITUAL CREATION – THE ULTIMATE QUESTION

Many of you, like myself, have walked a divided life path: Spiritual and material or scientific. The dichotomy has not been a bad thing. Faith and belief have given me peace, purpose and meaning. Worldly secular education and its application have provided me with adequate creature comforts, satisfying companions, friends and recognition.

The majority I suspect will accept this state with thankfulness, I did for many years. Some though, especially late in life, wonder "What is the ultimate reason for having lived?" Listen to the words of Aristotle, paraphrased here, as he drank the court ordered poison hemlock, "I go now but who knows if to a better fate or not." Others as well: Mother Theresa confessing her doubts, even Christ on the cross leaving us wondering as he says "My God, my God, why hast Thou forsaken Me?"[12] And what about Einstein? (Believing in a creator but not a personal God.)

[12] Matthew 27: 46

In my personal experience there have been moments, even years walking on one side of the fence or the other, finding at times seemingly conclusive proof that science explained everything without a God, but then later only to find more doubts. At other times finding proof positive of the existence of God, only later having doubts again. The endless loop of frustration follows the ardent thinker who pursues knowledge and eschews blind faith as the answer. I believe that both a rational belief system and a strong faith basis is preferred. Ultimately though, what about you?

The truth is that many, if not most of us, zigzag on this path. Some have lived exclusively on either the Spiritual or nonspiritual path.

Some begin their life in a home where faith is a core part of the person's character, where they acquire knowledge and life experiences, while others claim knowledge that disproves God. The sensibilities of the former may be shocked to learn that some people even flaunt a belief there is no God. Often this presents a serious personal challenge or struggle for them as they are unprepared to debate them. However, most are taught there are non-believers, and that as believers they should pray for them.

There are also those raised in homes where faith is not particularly discussed or employed. They may base their character purely on experiences and life events and then be surprised that people talk about belief in some mysterious God. It is more difficult for them to believe but when they do convert, they usually are staunch believers. Faith, like anything of value, will generally have more personal value if one has worked to acquire it. Think C.S. Lewis.

The fruit of knowledge has been taken by humankind to boost our own egos, some to disprove the existence of a creator and others to defend and prove the existence of a creator. But to each of us curiosity has a companion named doubt. Almost by definition a seeker is driven by doubt and curiosity to drive the active brain.

Many people have personal experiences which they believe to be divine intervention. Skeptics write the incident off as coincidence. Others say secondhand miracles are worthless. I will share some stories that you may find compelling. These miracles have been shared by a variety of persons, each with the hope readers will be touched sufficiently to investigate the reality of God. They are trusted colleagues, friends, and acquaintances.

Growing older and more observant of organized religion, I found plenty of things worth complaint, yet I thrived in the setting. I even wrote and delivered a sermon on youth pastor day. At the invitation four of the youth present joined the church.

At certain times in our scientific development, we stumble across discoveries so profound we believe them to be the ultimate findings. One example was the theorizing of, and finding, the Higgs Boson particle. It is even now often referred to as the "God particle" signifying it to be the ultimate answer. Of course, it was not long before other discoveries were made as well and while magnificent and beautiful discoveries are being found, nothing yet is being proclaimed as having discovered God. As a matter of fact, most of what I read is not mentioning science attempting to find God.

Many people have personal or anecdotal stories about their encounter with divine guidance and tell it so well that the impact is profound. One such incident came to me as I saw a patient in the emergency room. I will share it with you as a stand-alone chapter.

EMERGENCY ROOM MIRACLE–MARTHA

The sad faced woman's pleading eyes caught mine the instant I stepped inside the examining room. I do not remember what happened during the next few steps. Everything seems nonexistent between that moment and the moment of realization that my stethoscope was on her chest. Something miraculous had happened. Knowing the prelude to this event you might understand better how it fits into the narrative's totality of meaning.

My beeper sounded an unwelcome intrusion while I was at a Seekers small group. All faces turned to me. Of course, I should have had the beeper on silent.

"Dr. Logue, the emergency room clerk spoke authoritatively, "We are awfully busy. Dr. Roberts (the E.R. chief M.D.) needs help. We have a lady who looks sick. He thinks she will need admitting. Can you come and evaluate her and admit her if needed? We could use the help."

The Hippocratic Oath always bears heavily on a doctor at times like this, for it is at the core of a caring doctor's desire to help. I said "Of course, I'll be there shortly."

Nevertheless, mixed feelings rode with me to the hospital. They were being kind to me, a young doctor, to give me the opportunity of an admission. Unbeknownst to them, they were calling me out of an important meeting where I was truly seeking the meaning of life.

When I learned the patient had a chief complaint of having a cold for over a week it seemed even more inconsiderate that I should be interrupted this way when an office referral for the next day would have been more appropriate. Although I had obtained some spiritual uplifting I was called out prior to the social interaction. Mingling and personal interactive thought exchange is an important part of learning. I remember thinking "I'll take a deep breath and be professional,"... until I actually entered the room.

As mentioned earlier, I only remember stepping into the room and her sad, pleading eyes locking onto mine.

My usual routine when approaching a patient is to have a conversation, to establish rapport. I would then listen to the back of their chest, examine the neck, face, the eyes and then the front of the chest. In this case, this never happened. Without even realizing what I was doing I walked directly to her placing the stethoscope in my ears and hand. Reaching her I placed the stethoscope over her heart without conscious awareness of my action. Suddenly I was shocked to essentially come out of a trancelike state and realize I was listening to a patient's heart. Moreover, I was listening to a repetitive squeak-squeak, the telltale sign of pericarditis, an infection of the pericardium, a protective sheath around the heart. This is an infection that can be fatal. It is not common, and in fact it is somewhat uncommon. It certainly is not a diagnosis you would expect when someone approaches you with a chief complaint of a bad cold.

Completely autonomously my hand had taken the stethoscope and placed its diaphragm to a specific point on her chest as an arrow perfectly aimed, I know because I discovered by moving the stethoscope, even slightly, either direction, I could not hear the squeak. It was the only spot where it could be heard. Yet, neither one of us had spoken a word.

That event humbled me instantly, dramatically, affecting me to this day. I knew at that moment that my favorite prayer "God, if you are real and you are there, please show me," was being answered directly and profoundly.

Even then despite my prayers I was not sure I could fully accept these signs as absolute truth—or if I wanted them to be truth. This was one of the beginning signs that became increasingly undeniable, graceful, show-and-tell miracles in my life.

The patient and I had a thorough conversation about her diagnosis and treatment. I was able to start treatment, admit her to the hospital and consult a cardiologist the following day. She did well but had she gone home with simply being treated for a cold she quite likely could have died prior to returning for more treatment.

Instances like this are easily accepted by people grounded in faith. If they are not so easily accepted by those of us grounded in science we must all consider the odds and let the odds build up so that even a scientist can begin to see the odds become so overwhelming to many of us that it is pointless to resist further. The bedrock of our scientific foundation is when facts are logically aligned, we must accept.

But despite this seemingly unmistakable show of presence, dominance and control, I needed more. I exited that scene and wondered if that undeniable guidance was in fact from a Supreme power. My reluctance was surely known, and I wondered if a God that obliging would have the patience and grace to repeatedly show Himself?

I sensed the battle raging in my soul and cringed at my resistance to His persistence.

ANOTHER EMERGENCY ROOM MIRACLE

As a young doctor I watched a miracle unfold which astounded me—and humbled three much more experienced, first rate doctors.

After not being able to confirm or deny the existence of God, and maintaining my agnostic attitude, an event happened that certainly could be called a miracle. But would I be ready to accept it? I had just attended a small group of seekers of spiritual truth, when at the conclusion of the discussion, I received a call from the University Hospital emergency room. Because of their overload they asked that I come and see a patient for evaluation and treatment.

On arriving at the emergency room, I reviewed the chart looking for something of substance. It was evening and I had driven in just to see this man who apparently, reading from the chart, had simply rolled out of bed, had no pain or difficulty and did not wish to be there. He had no symptoms, but his wife thought he needed to be medically checked and had insisted (he emphasized the word with force) he come to the emergency room for an evaluation.

No matter what my preconceived thoughts might be to the provided initial information, when I am in the presence of a patient, he or she has my full, concentrated attention. This is generally the best way to get the correct diagnosis. As on this occasion, sometimes that process will be aborted by an unexpected shortcut. It cannot be explained by anything scientific that we know of. Even outlandish conjectures are lame. For example, a colleague listened to this story and observed that it was just a coincidence, and that I had been astute enough to make the diagnosis. Really? That does not hold up under reasonable analysis, as we will see later. This one is straightforward, and yet it is complicated and puzzling. Read on to judge for yourself – was this just a coincidence, or divinely guided event – a miracle?

Allow me to accurately describe the setting. From the hallway I entered the examination room through a door positioned in the middle of the hallway wall. The room was rather large. The patient was at a 45-degree angle to my right lying on a gurney pushed tight against the walls. He was lying on his right side with his head propped on his right hand and arm. He appeared to be comfortable and in no distress. His feet were pointing to my right; therefore, he was facing me. His wife sat in a chair at the head of the gurney. His back was snug against the wall. As I approached him, he began pushing himself up so that his head started to come up and his right leg stayed straight while his left leg bent and came forward over his right leg. We made eye contact as I began walking toward him, while simultaneously stretching my arm and hand out toward him requesting that he remain lying flat. As I did so he had already halfway pulled himself up just as I touched his abdomen with my right hand just below his left rib cage. All of this seemingly a very natural thing to do. The most unusual part of it was when I touched his abdomen, I felt a mass under his left rib cage. A mass was not supposed to be there.

He was simultaneously telling me that he felt great, and that there was nothing wrong with him. He didn't need to be there, and he wanted to go home. I asked him to please lie down and be still while I examined him. He resisted but his wife insisted that he cooperate, and he did. For me, the evening had turned from a question of "Why am I here?" to "How did I get here?" and "This is a terribly dangerous situation." Actually, I thought "this cannot be happening! I am in my first year of family practice and I was told I would never see this type of injury." With him lying flat on his back the mass was not palpable. It likely would not be reached even pushing hard under the rib cage to find it. Ordinarily, the mass would not have been discovered. If not for his unusual body positioning at the exact moment I placed my hand on the exact spot, it would have been missed. Though swollen to the size of an orange, it was not tender. He did not feel me touching it. He was totally unaware of its presence.

He complained that it was silly for him to be there, taking up everybody's time and space and he continued wanting to go home.

Again his wife told him to let the doctor finish the examination. He complied.

Because of the unusualness of this situation, I quizzed both the patient and his wife extensively. Nothing else was of relevance. I requested an x-ray which he did not want and refused but allowed after his wife's insistence. It showed an orange-size white object in the exact spot where the small spleen should be. These x-rays were before the automated x-ray machines and so were read while dripping wet.

This confirmed for me a highly dangerous situation. I was sure that this patient had a subcapsular ruptured spleen. This means the inside of the spleen was fractured and bleeding, but the tough capsule of the spleen was so far keeping it from rupturing. To my thinking it was only a matter of time before it would rupture and when it did, he would bleed, hemorrhaging massively and quickly. [13]

---

[13] In prehistoric times the spleen played a larger role in our survival because it acted as a self-contained blood bank for humans when they were lacerated somehow and bleeding. Being situated at the top of the abdomen under the rib cage it is one of the most protected areas of the body.

I fully explained to the patient and his wife my diagnosis, the danger of his situation and that immediate attention was needed by a surgeon. He did not want to do that. Again, his wife persuaded him. I explained that if he went home and it ruptured, he would be dead before he could get to the hospital. There were no such things as emergency home services in those days.

There were three reputable and well-respected surgeons in the emergency room that evening. All of them knew me and had a good rapport with me. However, one by one, as they heard my story, each one shook his head and said, "No, that can't be. It must be an artifact on the x-ray. It cannot be. There is no danger. Send him home. I don't know what it is, but it is not a subcapsular ruptured spleen." One by one, they all told me the same thing, each one looking at the x-ray, refusing to give any credence to my diagnosis and refusing to examine the patient.

With the help of his wife, I was able to get him admitted to the hospital.

After considerable persuasion I was able to get the lab to crossmatch six units of blood. I convinced the nurses on the floor where he was admitted that this was a dangerous situation and he could die during the night, that they should check his blood pressure every 15 minutes and immediately notify me if there was any change. At around 5:30 the next morning the nurse called to tell me, (swearing that they had checked his pressure every 15 minutes all night and it had been perfect and he had no difficulty), though at this last check he had had no blood pressure. They could not feel his pulse. Though he was breathing he was cold and clammy. I instructed them to get him ready for surgery. We already had an IV running. A blood transfusion was started. I called the switchboard operator and found which surgeons were in the house and talked to one that had not been there the night before. He immediately put this patient ahead of his scheduled patients and did emergency surgery and saved the man's life.

Background information is relevant. In medical school our professor of surgery practically apologized for telling us about diagnosing a subcapsular ruptured spleen.

He started with, "We are obligated to tell you this, but none of you are likely to ever see one." He practically invited us not to listen. After instructing us on a subcapsular spleen rupture he gave the following advice. "The spleen and the heart are the most protected organs. The spleen is protected by the rib cage. It takes tremendous trauma to get to the spleen. It floats so it is moveable. For example, it will not be trapped against a solid spine. Its capsule is tough. If it ruptures the person can be dead before they reach medical assistance. So, if an accident victim gets to you alive, perhaps think about this possibility if they were in a head on collision or fell out of a five-story building." Essentially saying, "otherwise, don't bother."

In this case, how could it have ruptured with this slight trauma? Why was he on his right side against that wall? If he had been on his left side, I would not have found it. In fact, if the timing of touching the exact spot at the exact moment he had unwittingly positioned himself where it would be felt had been two seconds off either way I would never have felt that spleen even doing a thorough exam. Some have said I was just lucky and smart. He was lucky. I say an intervening God saved a man's life and worked to show Himself—again—to a doubting soul.

HOW MANY LIVES?

The thoughts inspiring this chapter have not occurred previously, but now seem very important and relevant in the context of the book. Suddenly I am realizing that I have escaped death more than once. I haven't yet counted, but I will attempt to record the events here, chronologically.

The first two I've written about in more detail in earlier chapters. 1) I was born almost too healthy at 12 pounds 13 ounces. That could have killed me and my mother. 2) I related the story of my near death at twenty-six months old. Afterwards, I remained skinny until I was 40 years old.

While still a youngster—and before penicillin was discovered—on two separate occasions I had severe boils of the carbuncle type. In other words, multiloculated staph infections, once on my knee and once on the back of my neck. In those days staph infections were often fatal. Dr. Vickers lanced, scraped and packed with iodoform gauze. I remember so vividly the boil on my neck where at his home on his front porch he sat me on the steps and with his left arm held my head down and with his right hand he lanced the boil and did the scraping and packing. Obviously, I survived both of those potential death threats as well.

As a teenager I came home late one evening and intending to frighten my younger sister I made noises at her door. My father, thinking I was a burglar or worse, raised the rifle intending to shoot through the door but my mother cautioned him that it could possibly be me.

After my second year of college my parents were in critical debt because of severe illness and medical expenses. I stayed out of college for 15 months working to pay off their debts. During those times I became disheartened and seriously considered taking out an insurance policy and "accidentally suiciding," for it appeared to me that my parents were much more deserving than I was, and it appeared that I was going to have a hard life. Finally, I decided to try one more year of college. (In hindsight, I credit God in His infinite wisdom and mercy, gently leading me toward the beautiful life He knew lay ahead.)

Another truly significant event is recounted in a later chapter regarding the "Dreamer." You will see that that was too specific to be anything other than divine protection.

In 1962, during my family practice days, I developed severe chest pain. It got worse during the day. Our in-house x-ray showed a spot in my left lung. The radiologist thought it was cancer and recommended surgery. I saw an internist who said he had seen one person with a similar lesion who lived. He likewise recommended surgery. After prayer I decided this was likely not malignant but rather infectious because I served in the National Guard and had been to Mississippi for a two week summer camp training where histoplasmosis is rampant. I was correct, I had histoplasmosis. If I had the surgery, it would have likely spread and been fatal.

Some time later I had an illness which debilitated me considerably: Epstein Barré Virus. Again, I considered suicide but did not have adequate provisions for my child. I didn't want to die; I was just too tired to live. But again, by the grace of God, I did.

Looking back, I can clearly see how the presence of God has been there all along, guiding, directing and protecting me. I am grateful for every instance where He showed his love for me.

PRAYER POWER

You likely have noticed that many of the miracles in this book are a direct answer to prayer.

When we are awake spiritually, we have knowledge certain of God's presence. Sometimes we are more consciously "in tune" than at other times. For example, one might feel more connected to the Divine during meditation or prayer, whether in a group setting or privately.

The "Meditation Group Miracle" that I shared is a perfect example of being in tune. But even in that case I was not in tune consciously until God intervened, calling my attention to the message before me. To me it demonstrates unmistakably an intimate interaction by God in our lives.

Another example of being in tune, perhaps in a lesser or more casual way, might be the story where an accident was avoided when a passenger going up the mountain road urgently yelled to the driver "Take that road!" This person later commented "Good thing I had that impulse," as opposed to commenting "Good thing for all of us God saw fit to intervene." I believe that this is a fine example of someone being used by God as a channel, even when the person is not completely tuned in to the fact that the impulse was divinely given.

Often you will hear people say, "that was a God thing," "that was miraculous," or "that was a miracle," without consciously realizing or thinking about the fact that God just intervened to give them an intimate miracle.

Miracles can be and often are personal and intimate. They are His love and care for us. More and more people are telling their miracle stories. We should do more praying than just hoping, and the increase in the telling of miracles is becoming a recognized good sign. Believers should all organize and collectively pray. One excellent way could be that each of us stops to pray for 10 to 15 minutes to pray at the noon hour. Imagine the effect of God's love and intervention constantly sweeping the earth, 'round the clock.

We should pray that the Holy Spirit will cover and energize believers in our nation and spread through the entire planet, acknowledging the sovereign omnipotence of God. We should give thanks for His love and grace, ask His assistance in reflecting spiritual living, have righteous behavior and work toward peace and goodwill to all.

I sincerely believe, if faithful people began collectively praying to God at the noon hour each day for a spiritual resurgence in our nation—and the world—people, clergy, state, national and global leaders all would be blessed. National leaders should suggest a noon prayer to God for our nation. He likely would reward us, the nation and the world. I believe each of the three branches of government should have a chaplain. The President's cabinet should host a prayer breakfast and propose a suggestion that all believers on a voluntary basis, begin a movement to each day at the noon hour pray for a spiritual resurgence asking God to bless us with thankfulness, unity, harmony, peace, as well as world peace and harmony. The chaplain of each branch should endorse it. This encouragement would need to be personal as it could not be government supported.

Two scientific studies published earlier proved the power of prayer. I see it repeated regularly in a local prayer ministry.

Study A: Two corn fields were planted and treated the same except one field had a religious group pray over it. But not the other. The prayed-over field was healthier and produced statistically significant more corn.

Study B: Mice were purebred and genetically equal. They were treated identically in every way, except one had a religious group pray over them but not the other. Both groups were given dangerous levels of radiation. The prayed over group continuously had better retained health and longevity.

Along this line we should live with the knowledge that the manner in which we live is like an ongoing open prayer—or not—and exudes a corresponding influence in and around us. You must not lose sight of this. Equally important, the company we keep is an influence on us. Even if we do not know they are emanating good or evil vibes, we are nevertheless accordingly impacted. So be conscious and aware of your energetic surroundings and take care to protect yourself.

ENVIRONMENTAL INFLUENCE

I saw a very interesting published book of an experiment. A standardized system was used to produce water crystals from a consistent grade of water. Identical conditions were employed with one exception. Assorted books were chosen and left near the water during the crystallization process. Beautifully designed crystals appeared in the water surrounded with books considered good—for example, The Bible. The water surrounded by books considered not good or bad yielded misshapen, awkward or even grotesquely shaped crystals. All of this was supposedly due to inanimate influence. We should expect even more influence from our human animated being. We must note the crystallization process could have been influenced by the books left in close proximity. However, it might be even more likely the influence came from the experimenter's unconscious or willful thought process. We cannot know for sure.

My considered belief is that we are His favorite creation, His chosen people – collectively. He lamented over free will hedonism. He sent His son to clarify the way to follow. He allowed the torture and cruel death for our salvation and still, as a whole, He shows adequate appreciation, love, and commitment to His cause, agenda and service. He has shown us that He is expecting the same back from us. He has such vast power at His disposal. We only need to believe, and prayerfully, earnestly ask for His presence and assistance. "Ask, and it shall be given you; seek, and ye shall find; knock, and it shall be opened unto you." Mathew 7:7 and Luke 11:9

SPIRITUAL TEAMWORK

In early 2020 a woman who was a member of a three-member musical band told me of a miracle that occurred within their membership. The other two members were men. All three were talented musicians, each played more than one instrument and each of them were singers. Also, each wrote songs. Unfortunately, one of the gentlemen had become a habitual user of alcohol. This escalated to the point he was not capable of performing to his capacity or talent. The ability of the band to continue was becoming strained. The user denied that it was that big of a problem, or that he needed to quit or slow down. His mother indicated to the band members that she feared he was hopeless, as she had watched him turn away every suggestion for help.

The woman told me that at one point she misplaced her telephone for three days, and was therefore unable to reach him. She and the other band member, who had been a longtime friend of the wayward one, both began a prayer vigil. They prayed diligently, faithfully, earnestly, hopefully and standing in belief for an answer.

After three days she was able to find her phone and eventually got in touch with him. She learned that three days earlier he had woken up with the distinct realization that he should go into rehab. Without consulting anyone he checked himself into a rehab facility and stayed for one month. On discharge from the facility, he maintained sobriety, was thankful for the blessing of sobriety and eagerly participated in A.A. meetings. He continues working on recovery and learning more about it.

He also felt joy upon regaining his musical interest. His talents, friends and fellowship was rekindled. He rejoined the band.

The is a combination of the power of prayer and the Holy Spirit making an active divine visit with imprint on the person in need. So powerful was this combination that the rehab worked. Typically the percentage of the first rehab working is quite low. When a miracle is involved the odds become much stronger, better, and last longer.

A BENIGN UNSETTLING BY THE HOLY SPIRIT

It happened to Vickie Jones, (not her real name) a 57-year-old lady who, after a life of struggles, misfortune and abuse, sought mental health help. She soon felt better enough to go to church. It did not turn out the way she had envisioned.

At the end of the regular service, the congregation was invited to a special opening of their unusually large baptismal pool. Those that had previously been baptized who wished to enter the pool and "get closer to the Holy Spirit" were invited to come and enter the pool. She quickly made her decision. This was not something she had wanted to do, but no sooner had she made the decision that she realized she was getting out of the pew and walking toward the pool. She tried to stop and turn around, but her legs kept walking. She could see nothing beside the path that she was on. It was as if there were no other pews or other people, just the path to the pool. Before she reached the Baptismal and even before the walk, she had the thought, I have already been Baptized. There is nothing more to gain by doing it again.

Reaching the pool, she did not want to go in but someone assisted her and she stepped right in. As soon as she did, she pulled both hands to her mouth and started screaming as loudly as she could, "Get me out of here." She did not want to be there.

At that point a very nice, caring gentleman asked, "Are you ready to let it go?" She had the perception that he knew about the sexual abuse she had endured, the mental, emotional and physical abuse from her first husband, having been through two bouts of cancer and a hard life in general. She indicated that she was ready. The gentleman then said to her "You have a fear of dying. You fear that death is the end and there is nothing else after death except darkness. You fear that death is painful. It is none of that. It is peaceful. You will be a point of light for others." Amazingly the fear left her as if it had never been there. She felt empowered to be normal and to be able to face anything that might come her way.

She was asked what she wanted out of the experience. She answered, "Peace." She decided that the thousand points of light she saw above her were collectively the Holy Spirit.

She was feeling much relieved and thinking it was time to go but the gentleman said he had one more thing to do for her. Reaching toward her chest he started pulling as if he were pulling things from her chest. It appeared that he pulled large objects out of her chest one by one until she felt emptied of her past traumas and abuses. He had told her that it would be painful, and it was. She could only equate it to having had an operation. She was weakened but she was healed.

Afterward she apologized to her friends about screaming. They told her she had never screamed. They had seen her put her hands to her mouth and had heard her whispering, but not screaming. She commented that it had felt as if she had been in the water for hours. Her friends told her she had only been there several minutes.

Vickie stated that the experience had been so draining that she had been mostly in bed recuperating for three days, before being able to get up and function at her normal level.

She knows this was the Holy Spirit taking the bad from her and the reason why she is now a new person, happy and confident of her salvation and her ability to face life normally.

CONSECRATED

A 76-year-old retired Methodist minister, a wonderful gentleman, granted me an interview during which the following miracle was related. His lovely wife sat in and it was a delight to see the joy on their faces as they shared the story with me.

During his ministry he had seen people in all kinds of conditions, but this miracle involved a member of his church who had cancer, and who had been gradually withering away. He had been summoned to come give her last communion.

On seeing her, she was emaciated, eaten away by the cancer, and too feeble to reach out and hold the cup. He remembered holding it to her lips so that she could sip the Sacrament.

Because of the fact that in recent times some churches do not consecrate the bread and wine, I asked, "was it consecrated?" He had an absolute look of shock on his face. "Of course it was," he said. A short discussion ensued. He and his wife both passionately believed in the need for the symbolic sacraments to be consecrated to be a real Sacrament. Many others bemoan the fact that people change important things for convenience, thereby unwittingly destroying the value of what they do.

He did pray with the lady but did not say that he specifically asked for healing. He strikes me as the type that would ask God to do His will.

The minister then told me that several days later he heard that she was still living. Her death had been expected the night he gave the communion. A few days later she was still alive and appearing stronger. She continued to progressively regain her strength, and her health improved to the point that was her normal self again.

When she later visited her physicians, through the examinations they determined that she had no further cancer. Indeed, she lived for another ten years cancer free, eventually dying of a heart attack.

After returning to church, she told the Minister that the moment that she took the sip from the cup she felt inside of her what she described as a pop. She said that it was at that moment she began to feel better.

Both the minister and his wife, knowing the woman well, recognize this as a true miracle.

As a result, some of the other parishioners who had witnessed this recovery began calling him the faith healer. Telling me this story he simply laughed. He did not see himself as a "faith healer," He saw himself as a man of God and man of faith, and God as the healer.

SEEK ME

Jeremiah 29:13 says, "And you will seek Me and find Me when you search for Me with all your heart."

We must remember that God has given us free will. Therefore, in general we should not expect Him to individually decide to block our free will. When we choose to not follow God, He allows that. It is when we choose to seek Him, He will assist us in finding Him.

My case is an example. Growing up being a believer was my natural state. However, observing the flaws in church members and ministers, and simultaneously being excited about the unfolding scientific developments, I was lured off the spiritual path.

He allowed me to pursue a journey on that alternate Doubting Thomas path. Along the way he presented minimal signs to me that I barely recognized as Him reaching out to me.

When I earnestly needed to know for sure, and asked His assistance to find Him, in other words to show me if He was there, He obliged in the several ways I recount in this book.

Without knowing the scripture, I had succumbed to the message from Jeremiah 29:13 and sought Him with all my heart.

DOUBLE DUTY MIRACLES

Friendships don't always divulge everything about each other. A couple I know, Patricia and Patrick, not their real names, and I, all being seniors, share some common interests. Recently in a conversation with Patricia she mentioned the importance of her spiritual beliefs. Being unaware of that aspect of their lives I mentioned to her that I was writing this book. Her immediate response was, "Let me tell you about our miracle." She then related that she and Patrick had shared their stories many times. Naturally, I was all ears to hear another miracle story.

Patricia and Patrick had been a loving married couple for several years, but Patrick had succumbed to drinking and then to its addiction. She related this as an exceedingly difficult time for her, and for him as well. Tolerating what alcohol does to a couple or family had come dangerously close to destroying their marriage.

In the fall of 1983, she relates, she went to work that day leaving Patrick at the house to meet a repairman. When she returned home that evening, as I recall the repairman had not come. Patrick had been drinking all day and was significantly intoxicated.

Patricia became furious. An argument ensued.

Patrick left the house and drove away. He went to a friend's house and drank more alcohol. Patricia fumed, thought up things she could say to him leading up to a divorce. She was so angry, nothing else mattered.

Patrick left his friend's house extremely intoxicated and had a wreck apparently after passing out. His car hit a guardrail and skidded along, scraping the side rail so far that it scraped almost all of the rubber off the tires of both driver's side wheels. He remembers at one point being awake and seeing sparks flying.

When the car finally stopped, the driver's side of the car was unmanageable. He had a difficult time getting out of the car but did.

Simultaneously Patricia was semi-praying, being mad with Patrick and mad with God. She was in a complete state of despair and determined that she would divorce him. Then it happened. A voice came to her plain and clear, "be still and know that I am God." This had a strange effect on her. She did not know what it meant. It surprised her but did help calm her.

At the same time Patrick had managed to get out of the car and shortly called her informing her that she needed to come get him. He had started to walk but fell on his hands and knees and began retching and vomiting in a horrible fashion. Then suddenly he was out of and above his body, watching the scene of his body retching and vomiting. Initially, he did not realize it was himself he was watching. His thought was actually, "That poor miserable soul is in terrible shape. He really needs help." That is when he realized he was watching himself.

Patricia drove and brought him home and they drove into the garage. She knew she needed help and so she summoned their 15-year-old son to come help get Patrick to bed. Patrick literally fell across the bed and was left there to sleep.

Patricia went to the den and stayed on the sofa with very restless, poor sleep that night. She was resolved she was going to divorce him. She decided she would tell him the following morning.

Morning came and Patrick walked into the den and walked to her and said "Patricia, I am sick. I need help. I am an alcoholic."

Eventually in telling their stories to each other they realized that God had spoken to her to be still and know that he was God. At the same time Patrick was having his out of body experience. He realized he had to have help.

The foregoing happened on a Friday. On Saturday they went to an A.A. meeting. A friend of theirs gave Patrick an A.A. book. When the members realized that it was only Patrick that had the drinking problem, they needed Patricia to leave, but miraculously they voted to let her stay. During that meeting they rebounded and resolved that he would go to A.A. meetings and she would go to Al-Anon meetings. They have both persisted in that to this very day.

It had been days after the incident before they had told their stories to each other and realized that God had done double duty and reached out to them, changing both their lives forever. They realized the personal interventions, the direct one-to-one power from God performing miracles simultaneously to two individuals who were miles apart physically, as well as the miracle to their marital relationship, creating between them multiple miracles.

A MINISTER'S SURPRISE

Another minister acquaintance of mine told me an interesting story. Like many ministers, my friend's sermon delivery appeal spread broadly. At the same time, many churches either needed or desired a minister to fill in on a Sunday they did not have one, or from time to time just desired to have an outside minister visit.

My friend on occasion was asked to come and bring the message for a church some two states distant. He frankly enjoyed opportunities to visit other congregations. On this occasion, his wife who was pregnant, went along with him. In retrospect, unsuspectingly, she would witness a miracle. Traveling was a good opportunity to be together—usually without interruptions. They looked forward to a joyous opportunity to have time together.

Two-thirds of the way to the church their car began to lose power. Worried, they stopped at a diner. The few servers they spoke with did not know of a mechanic they could call.

After a while, a gentleman came in, appearing to be a local, judging by the way he was dressed and his voice. He came to my friend and spoke words to this effect, "So, are you the gentleman with car troubles?" Assuming one of the other employees must have known someone, my friend answered, "yes." The gentleman offered to have a look with him.

Together they went outside chatting about cars and various things related thereto. The gentleman asked him to get inside and pop the hood, which he did.

He watched the man lean under the hood and not much later the gentleman asked him to try the car. It cranked instantly and was running perfectly normally. My friend was greatly relieved and thankful.

The gentleman told him he should have no further trouble.

They continued chatting as they went inside to wash their hands.

While washing his hands he turned to thank the gentleman, but he was not there. He had not heard him leave so he went to the door to look but did not see him. He asked where the gentleman had gone. The people unanimously told him they had not seen anyone but him come into the diner or go to the restroom. Asking his wife, she said that she hadn't known why he had gotten up and left, and like the others in the diner, she hadn't seen anyone with him.

Only then did the surprise of his life come when he realized it was an angel amongst us. He was caught completely off guard because this appeared in every way to be an ordinary person who showed up, magically fixed the car and disappeared. No one else saw him or had any explanation for what had happened.

The trip resumed in normal fashion. Both praised the Lord for help, especially in this very unusual and intimate way, where an angel appeared specifically to them to help them with their very specific need. They had no further problem with the car. He continues to feel awe and excitement when telling this story as his miracle.

HE SCARED ME

A fifty-six-year-old man basically interrupted a conversation between another person and me. He happened to think of a story his father had told him, and was so excited to tell it he began sharing without asking if it would be okay. He simply jumped into telling the story.

His father, long since passed from this world, in the latter part of his life told his son the following story. The son knew his father had been an alcoholic but was nevertheless a religious man who believed in God.

"Son", the father said, "something happened to me a few years back that I have to share with you. I went fishing that day, had all my equipment with me, bait, rod and reel, everything I needed. I went to a familiar spot at the river, set everything up and began fishing. No one else was there. The fish weren't biting but I kept fishing."

"Everything was quiet but then I heard some crackling in the woods just behind me. It sounded like small branches were being stepped on and broken. I turned and looked and there was a man standing there. He just stood still and looked at me. He never said a word. I went back to protect my things, and thought that I might have a conversation with him. But when I turned to approach him he was nowhere in sight. There was no trace of him. It was eerie."

"I decided to resume fishing, but I saw a rapidly approaching cloud. Everything was getting too weird. I had an uneasy feeling and did not understand what was happening. Fright overtook me, and so I gathered all of my things as quickly as possible, and I went home."

"The next day, I went back to resume fishing. What I saw on arrival gave me chill bumps. The spot, the tree that I had been sitting by had been struck by lightning. I knew then that an angel had come to me and had saved my life."

"Son, it made an impression. I cut way back on my drinking and as you noticed, eventually totally quit. That incident changed my life."

"I just figured you would want to know what happened to me that day."

A NULLIFIED MIRACLE

A man of my acquaintance, thirty-five years old confided the following to me. The quotation marks are for emphasis as every word may not be exact but truly close.

"My life has been a mess. I made it worse by becoming addicted to drugs and alcohol. I hated who I had become. My depression was dark and unbearable. I overdosed to kill myself. It did not work. I got worse and deeper in the hole. I bought a pistol, loaded it and looked at it often for when I needed it. The next attempt would be fool proof. Then life became too painful to continue. I sat with the loaded pistol three nights practicing and readying myself. On the third night it came to me 'This is it, no more procrastination—do it now.' My hand with the ready gun moved upward. The phone rang. It broke my thought and action. I answered. A lady with a distinct African American southern voice asked for someone other than me. With an unsteady voice I told her 'That's not me, you have the wrong number!' She said 'Man, something's terribly wrong with you. You need to get down on your knees and pray.' I did exactly that as if it were a command. After praying for a while every worry or problem I had been harboring vanished. I felt normal, and in fact wonderful for two weeks. I visited my ex-girlfriend to share the good news. I had my life back. We had a good visit. Unfortunately she had some drugs. We celebrated my good fortune with her convenient drugs."

"The next day all my worries and problems were back in full force."

This demonstrates the powerful grip of addiction. If the addictive substance is close at hand, it has a strong chance of snaring the addict. However, those tightly tethered to the Holy Spirit have a significantly better chance of resisting the temptation.

MISUNDERSTOOD MIRACLE

In 1962, while I was a medical student, a fellow student and I were invited to participate in a civil defense study in Athens, Georgia. The study consisted of 30 people who were randomly selected from the street to volunteer to come into an underground shelter, prepared as if it were protection from a nuclear bomb attack, and stay there for two weeks. The researchers needed to know what to expect from persons confined with meager facilities and supplies. For food, there was tightly compressed grain. We had highly hydrophilic and iodinated water, and very basic hygiene and toileting. No trips outside of the room were allowed. We were monitored with a camera and a one-way glass.

Prior to starting the two weeks, the two of us were advised that on their one previous study, most of the thirty people left shortly after beginning, and essentially they had received no valuable information. They needed us to be mindful of the need to have the information and of our responsibility to do all in our power to see that we do not lose the community.

The group consisted mostly of adult males from 20 to 40 or 50. Some were older, and some were children.

Within the first 48 hours one of the gentlemen who was there with his approximately five-year-old daughter insisted that this was more than the daughter could take and they requested to leave and were granted permission. A few days later a young boy around the age of ten happened to injure a wart that was on his hand which began bleeding. He immediately wanted to leave. His father gave me permission to talk with him privately over in a corner.

The boy went with me willingly. We chatted some about the wart and that he would not need to worry about it. I talked with him gently and relentlessly, guiding him to a hypnotic state, finally in a deep hypnotic plane five. In that state I was able to advise him that he did not need to worry about the wart because it was going to die, turn black and fall off and it would not bother him anymore. Furthermore, he would be able to feel good, happy and enjoy himself as an adventurer. On awakening he was in good spirits, had no desire to leave and did not worry further about the wart, which did turn black, die and fall off.

Using personal skills, the two of us were able to maintain a high percentage of the initial group allowing the experimenters to gather considerable information.

I found it interesting that the first debriefing question was directed to me, asking how I made that wart die and fall off, and keep that child there.

Years later, after an internship and five years of family practice, I returned for a three-year psychiatric residency which had been my goal. Residents, like any group, have times when they chat about various aspects of the human condition. On one such occasion a few of the residents and I were in conversation when I told them the story about the wart. They laughed and thought it was smart to know that the injured wart might die and fall off. But I advised that is not what happened. What happened was the hypnotic suggestion let the brain reprogram the body's defenses to kill the wart so that it would fall off. They chided me that it was impossible for hypnosis to make warts go away.

I challenged them to find someone with warts and bring them to me and we would see. Within days they had found a lady with both hands literally covered with warts. Talking with her privately she would do most anything to get rid of the warts, so the hypnosis was a grand idea for her. I explained to her how I wanted to proceed, and she agreed.

I performed the hypnotic induction and she was a plane 5 subject which means she was an exceptionally good hypnotic subject. At that point I asked the group which hand they wanted the warts to be removed. All said both hands.

Being an experimenter by nature, I told them that would not work because if they all went away, they would call it coincidence. But, if we did it one hand at the time, they would have no choice except to believe that it was the hypnosis. They agreed and they chose a hand. We gave her the posthypnotic suggestion that her body would assess the situation and would cut off life sustaining supply to each of the warts on that hand and they would subsequently die, and then normal skin would regenerate to a normal hand. At this time, I cannot remember whether it was two weeks or four weeks when she came back. When she did come back, however, that hand was as clear as anyone's hand. You could not tell where any wart had ever been.

The residents were impressed but still not totally convinced and wanted to see the other hand treated, which we did, and which also worked just as well. They had no choice except to believe. We had all heard stories where people have bought a wart from a child for a penny, or other such techniques where a person has been given reason to believe the wart would leave, and it did.

Some would call this a miracle. That would be correct. But they would be incorrect in the nature of the miracle. We attribute most miracles directly to divine intervention. In this case that is not the situation.

The true miracle of this is that God has implanted in us a self-healing power. It has been used in many ways. Hypnosis is one way. It focuses the forces available to the body so intently that it can increase or decrease blood flow, increase or decrease body temperature, and even increase or decrease urinary output on their own. I have, for example, been able to demonstrate taking a 22-gauge needle, pulling the skin of the arm up and then pushing the needle through both layers of the skin and withdrawing it, with not a single hint of blood coming out on the needle. Nor could it be seen where the needles went in, or out. This power of self-healing is a miracle we use too little.

I once delivered a baby using hypnosis when I was in family practice. That baby was delivered healthier and more vibrant than most—measuring a 10 on the Apgar score, which is the highest health rank a baby can have on delivery. I believe it was because no drugs were involved.

A similar example happened to me on another recent occasion. Approximately three months ago in church one Sunday my right wrist began to itch. I started to scratch it as it had itched several times during the past week. When I looked at it I noticed it was actually a skin cancer, deep enough I could have set a green pea on it and rocked my hand and it would not have rolled off. My initial thought was, I'll need to see a dermatologist for this. My next thought was, 'Well, I'm in church, perhaps I should have a healing prayer, and have the prayer take care of it rather than a dermatologist.' So, it being prior to the service, I said prayers for healing. By Wednesday of that week, it was half gone and by Saturday of that week, one could not tell it had ever been there.

The more faith we have the more we can control our health. This doesn't mean that we have complete control of our destiny, but it is something that we (and science) have noticed repeatedly, and none of us should ignore the opportunity to be an active participant in our own healing, through faith and prayer. Prayer is a form of meditation for our own good, for our own use, but it also is a plea to God, the Holy Spirit, Christ and for those in other religions as well.

ANGELS AMONG US

There have been stories of angels appearing throughout history, as far back as Bible times and even earlier. Often, we don't get the opportunity to hear these stories firsthand. The previous chapters recounted a few such miracles, all of which were given to me firsthand. Here I share a few more, and include a couple mentioned in the Bible as well. In each case, there appears to have been angelic or divine intervention, producing results that cannot be explained in any other way. Let's recap with a bit of analysis.

A minister's car breaks down, but no mechanic is available. A stranger walks into the diner, comes directly to him and approaches him with the words to the effect, "sir, I believe you need some help with your car." He is attired properly for the area. His language and conversation are appropriate to the current situation. They talk casually about cars. He happily goes to the car with the gentleman. The minister is asked to get inside and pop the hood which he did. The stranger opens the hood and performs some work. Very soon he asked the owner to crank the car. The car then runs fine. In fact, the minister is told he should have no further difficulty. Back inside the man totally disappears and not a single other person has seen this helper.

A little girl five years old is extremely sick, and too weak to get out of bed. Her mother lifts the child up as if to heaven and pleads for the child to either go to heaven or to be healed. She places the child in bed and leaves the room walking down the hall. The child comes running out of the room shortly thereafter exclaiming that Jesus has come to see her. She told how He picked up a teddy bear and gave it to her. He tells her that she will be healed and well, then He disappears. Thereafter the child was well, and showed no signs of residual illness.

A three-year-old boy is about to fall from an escalator into an open space of two stories. A woman previously not in their presence appears, catches the child just as he begins to fall, places him down safely and then she disappears.

A Doctor of Theology, also a minister, was on a missionary trip to a third world country and was there during a time of dire community crisis with no solution in sight. An appropriately dressed and current language speaking man happens through, holds discussions about the crisis, and in about a day the situation is resolved. The crisis and the man disappear. Despite major efforts to find this person's identity, it is as if he never existed.

One of the most amazing miracles recorded in the Bible is that Christ was crucified, dead and buried. We know that he descended into hell for a period of time, only to return. We never considered that his body went to hell, so it was his spiritual body that went to hell. I have heard several persons, even a couple of ministers comment that if his body were not physically risen then Christianity is a fraud.

Another astounding Biblical example to consider for an even more complete understanding of this power would be the total revitalization of Sarah's reproductive systems at the advanced age of ninety, so that she would have her son Isaac.

These examples demonstrate the awesome power of God, and I believe they prove the existence of angels, intervening to help and protect us. Examining the precise timing, the perfection of how they were able to help, and the fact that people attested to them appearing and disappearing before and after the need was met, who could argue otherwise? It's also noteworthy that in each instance these "angel people" looked and spoke in every regard exactly like the local people—rather than being a shiny being with wings dressed in a white gown, for example. Surely God, being that powerful and resourceful, would also be able to raise His Son from the dead. How preposterous a divide in logic would it be to believe otherwise?

* Do not neglect to show hospitality to strangers, for thereby some have entertained angels unawares. ~ Hebrews 13:2

RECENT ANGELS

Hearing firsthand personal accounts of miracles with angels appearing in our presence should give us comfort in God's awareness, compassion and awesome power to provide instantaneous solutions.

Two such miracles come to mind. Eric Metaxas in his book on Miracles[14] discusses a scene where a small group of teenage girls bodily threw a girl they disliked over the edge of an overpass where traffic streamed below. Miraculously, an angel instantly appeared, caught the girl, brought her back over the side wall and placed her safely on the overpass.

Miracles such as this are spectacular and difficult to grasp. Atheists will proclaim, "a second-hand miracle is worthless," meaning that if you didn't witness it yourself, you can't give it credence. You can be assured, however, that these miracles were immensely powerful to the firsthand observers.

---

[14] Miracles: What They Are, Why They Happen, and How They Can Change Your Life

I encountered one lady with a similar story. She was at a local mall with her son, who was three years old at the time. She also had her one-year-old daughter in her arms, plus a stroller and bag of infant necessities and other shopping items.

They stepped onto the escalator going up. She noticed that her son had put both hands on the moving handrail and leaned into it allowing it to pull him up the escalator. She thought nothing about it. At the very top she noticed on a glance that he had not released his grip. He was being pulled forward and over so that he was about to fall into the space between the up and down escalators—some two or three stories high. She was helpless and horrifically frightened. Miraculously, a woman appeared and grabbed her son just as he started falling, placed him back in a safe area and vanished. The lady and several bystanders who had witnessed the incident tried desperately to find the woman. No-one had seen her before the incident. They saw her appear, grab the boy, return him to safety and then disappear. No-one saw her after.

This mother—all these years later—continues to tell the story passionately. There is no doubt in her mind this was an instant angel who momentarily appeared, saved her son's life, and disappeared.

Another acquaintance told me of a miracle in her life. When her daughter was five years old, she became quite sick with intestinal problems. Though she was otherwise in good health, she could hold nothing down, so she continued to deteriorate. Even hospitalized and being treated, she continued to decline. She was losing weight, strength, and alertness rapidly. The situation was desperate. The doctors were puzzled.

The child's mother was extremely stressed with the situation. Not knowing what was wrong or what might happen, she feared the worst. In a moment of desperation, she lifted her daughter off the bed, held her in both her arms, extended them forward and prayerfully said "Jesus, if You really need Claire and You want her You may take her. She is yours and please take care of her. However, if You don't really need her, I do, and I want her. If it can be Your will, please heal her and leave her with me." She then returned her child to the bed, limp and unmoving. With a sad heart the mother walked out of the room and down the hallway.

When she was near the end of the hallway, she heard her daughter calling for her. She turned and looked. Her daughter was up and running toward her. When she was closer, she said, "Momma, Jesus healed me. I'm well." The mother claims that she looked well. They discussed the scene further. The mother asked Claire how she knew it was Jesus. Claire said that Jesus had walked in the open door and came to her bed, picked up a teddy bear and gave it to her. He told her she was going to be healed. She would be well. Claire said she recognized Him because He was wearing a white robe, sandals, and long hair. She had been instantly healed with no residual problems.

Did you notice, the teddy bear added authenticity to a physical presence. Further to the power of the miracle the child was instantly fully well though she had been weak and incapacitated only moments before. How could we disbelieve this with so many ancillary validating aspects?

THE MIRACLE LADY TIMES THREE

The enormous lady sat in her small cubicle, designated as a receptionist room. With her one good eye she watched me approach. I could not help but notice the other eye cocked awkwardly up and away. My nonchalant greeting correctly conveyed my oblivion to even a suspicion that this sick woman would in the next few moments become the most important person in this chapter of my life. Without her I could never have written this book.

"Dr. Logue," she said, "do you take new patients?"

Though I recognized the compliment, I trembled at the idea of me treating and being responsible for this lady's medical care and health. I am sure that I paled at the thought.

"Adele" (not her real name) "you are seeing one of the best internists in town. I am a general practitioner. I have only been in practice one year and plan on going back to specialize after four more years. She interrupted me, "Dammit, if you don't want to treat me, just tell me. I am a big girl."

It might have been the implied hook 'you are supposed to help people, you can see I need help and you are turning me down.'

I did feel sorry for her and I would like to help. I knew she was often off work because of her illnesses.

She spoke again "I know how you treat your patients; I see how you care. If you take me, I will be your best patient. I will do everything you say."

My resistance melted. I offered to accept her as a patient. We agreed that we each would do our best. I likely had more doubt than she, and she more faith than I. On her first visit to the office, she listed one by one her numerous illnesses being all together more than a dozen average patients. Sitting on my examination table she was obviously oversized for the table just as she was with the chair in her reception area which was barely larger than her.

She listed her illnesses quite matter-of-factly. My diabetes is out of control, my hypertension is fickle and unpredictable. I have had a stroke. That is why my right eye is cocked up to the right. I have recurrent unstable severe angina, and nothing relieves it except I.V. Demerol and Phenergan. I have a chronic urinary tract infection with proteus. The antibiotics won't cure it. It invaded my left breast and they had to amputate it. I have had three heart attacks. I think the angina is pushing a fourth.

"Adele, how much do you weigh?" I asked.

"Dr. Logue, I have to go to the meat market to weigh. It is embarrassing, the last time I went I weighed 480 pounds."

'How can she climb steps?' I wondered.

For two years, we each upheld our contract. She did everything I asked, and I did everything I knew to do. As often as twice per week I would wind up at her home somewhere between 11:30 PM and 4:00 AM administering IV Demerol and Phenergan for her severe angina. She had lost forty pounds in two years.

The relatively small community of doctors in private practice collectively enjoyed teasing me asking when I was going to get Adele to lose weight.

It was a tough two years. And at times, just as it was prior to becoming my patient, she would periodically require hospitalization. However, she would not give up.

During one of Adele's visits she commented, "Dr. Logue, I know the other doctors are ragging you about me not losing weight because they are me. I am so sorry, but you know I am trying." She was right on both counts. She had lost barely forty pounds (8.33%) of her weight. Despair stalked both of us.

As fate would have it , around that time I attended a bariatric seminar. It was an 'out of the mainstream' aggressive medication program. And counter intuitively it required normal eating.

Adele balked, “I said I would do anything you told me to do but if I eat three meals per day, I will gain weight. I won’t even be able to get through the door.” She did, however, accede to her promise to do as I said. She followed the diet. She was compliant in the medication and eating, stating she would try it one month. At the end of the month, she walked into the office smiling and happy. She said she felt much better, healthier. She stayed with the program and steadily lost weight and gained health. Over the next two years she lost two hundred pounds winding up exactly half of where she was when we started. I felt she had earned a reward so as I gave her a bathroom scale which ranged from zero to two hundred fifty pounds,

During those two years Adele was healthy. She not only improved her life, but she improved mine. I never had to get up during the night to treat her angina. It had gone away. Her diabetes came into control. Her pain let up. She never needed hospitalizations except for one incident which is described elsewhere. She called me early one morning saying she had to go to the hospital but that she did not know why. (To be continued in the next chapter, “The Dream Gift.”)

THE DREAM GIFT

Fran, my efficient but pleasant receptionist answered a strange phone call from Adele "Fran, tell Dr. Logue I need to go to the hospital."

"I am sorry Adele, what's wrong."

"I don't know. I just know I have to be in the hospital."

"Are you having chest pain, or feeling sick?"

"No, but I have to go to the hospital."

"Let me get the nurse for you."

"Okay."

"Hi Adele, tell me what is wrong," nurse Florence asked kindly.

"It's the oddest thing Florence. I don't know, but something is wrong. I just know it but I don't know what it is. I am scared. Tell him I have to go in the hospital, and I don't want to go through the emergency room."

"I'll talk to him and call you back."

"Dr. Logue, Adele called. She wants to go to the hospital."

"What is wrong with her?"

"She is scared for one thing, so she is anxious."

Florence gave me all the details. My thought process was something along these lines: "I listened, I spoke, I had never known Adele to lie, therefore I know that she believes what she is saying. If I do not admit her, she will think I don't believe her. If something happens to her, how could I forgive myself?"

I admitted her and gave orders just as I had in the past. My evaluation of her revealed nothing more than what I had learned from the phone conversation and Adele offered an apology. "I know how strange this is, but I am scared. I just know I have to be here."

I was puzzled but decided that this had to be an anxiety state for some reason unknown to me or her. It should not be a big deal and should dissipate within a day or so. I made sure her usual tranquilizers and narcotics were ordered.

In retrospect it appears that both of us were conscripted actors by some unknown force. Could it possibly be an extraterrestrial author providing yet another human saga of such incalculable odds that no one can deny Divine intervention? Some of us—in a league with Thomas—can be stubborn in appropriately crediting such events.

The following morning Fran answered the phone shortly after 9:00 AM. Adele's voice was urgent "Fran, tell Dr. Logue I can't wait until tonight to see him. He has to come see me today."

"Why, what's wrong?"

"I have to talk with him about something that can't wait."

"I will get Florence."

Florence and Adele repeated the verbal dance. Florence filled me in.

"I'll see her tonight" was my response, admittedly spoken irritably.

Florence relayed my message but without the emotion.

Adele demanded "It may be too late. I don't know what it means but it is important. I am scared and I really need to see him."

I was determined not to be sucked in by her unrealistic demands. However, throughout the morning I couldn't shake the insistent urgency Adele kept pushing. I skipped lunch and drove across town to visit her in the hospital.

Adele sat upright in the bed. She spoke lively with the nurse in her room. Nothing appeared to be wrong. When she saw me enter the room her entire demeanor and expression changed. I witnessed how anxiety instantly gripped her face.

"Thank the Lord you came." Motioning to the nurse to leave "I have to talk to my doctor privately."

"Dr. Logue, I don't know why this happened to me. Nothing like it ever happened before. It scared me something terrible. I knew you had to know, and I had to be the one to tell you."

"What Adele?"

"I had a dream last night. You and Dr. Solomon were at a birthday party. The guest of honor was a woman with dark hair and olive complexion. The party was at a house with an all-glass front. There was one tree in the front yard. It was raining. Nineteen people came to the party including the guest. During the party, the doorbell rang. Both you and Dr. Solomon went to the door and opened it. It was awful Dr. Logue. A man with a shotgun was standing there. I saw him shoot you both and watched both of you fall. I woke up horrified. I knew I had to tell you, but I would have to wait until morning. I could not go back to sleep. The nurse asked what was wrong. I sort of lied. I told her I was having a panic attack. She gave me a tranquilizer."

"Adele" I began reassuring her, "you are correct to tell me, but you should not worry about dreams."

"But it was so real."

"I am sure it was; I see your emotions. I'll be alright. There is nothing to worry about." Thanking her I rose to leave.

"Please don't go, that is not all. The tranquilizer took effect and I went back to sleep. Dr. Logue, I had the same dream with the identical details. I have never been so upset. Sleep was impossible. I told the nurse I was having bad chest pain. She gave me Demerol and Phenergan. It knocked me out. Since it had been so long since I had needed it. Then it happened again, the same identical dream. It has got me really scared. I don't know if it means anything or not, but I think you need to be careful."

Thank you, I will Adele, but you put this out of your mind. You do not need to worry about it, or me." She was relieved to have told me, but she was not reassured as to my safety.

To my knowledge she had no way of knowing that Dr. Solomon and I were friends or that I would in fact be at his home that evening for a birthday party. Glass windows were the entirety of the front of their house except for the door. I doubt that she would have known that there was only one tree in their front yard. She could not have known it would be raining. The guest of honor was a midwife from England. She was Persian by birth (with dark hair and an olive complexion), which Adele could not have known.

There were just too many coincidences to ignore what Adele had shared with me. Add to that the fact that the exact same dream had come to her 3 times, even under medication. Could it possibly be prophetic? Back at the office I called Dr. Solomon's wife Naomi, and relayed the strange occurrences. Initially after hearing about the dreams, she assured me I need not worry, but shortly afterward she called back to inform me that the location for the meeting was being changed to Ann's house which was across town.

When I asked "Why?" Naomi's only response was, "We think it is prudent."

I knew there had to be more. In fact, there was much more at play that I was unaware of at that time. It was revealed to me over the months and years to come just how significant this message from Adele had been. I now believe it was actually life saving for me and for Dr. Solomon.

THE DREAM PROPHECY UNDERSCORED

This was a warning prophecy, presented through the dream world, and delivered to us through unnatural means from the 95.1% dark matter side. It was written, orchestrated and produced by a genius author—which could be no other than God—communicating with our side from His side. Those who might attempt to refute this will labor in vain for there is, and can be, no other suitable explanation. Interestingly, this scenario happened during my "doubting Thomas" days, but while I was genuinely seeking answers.

Let's take a look at the overall backdrop to this miracle:

The patient's condition had improved dramatically, from poor health with multiple life-threatening problems and several high risk factors, to marvelous health with joy of life and work. She had had no hospitalizations in many months.

Futile treatment of great intensity, including frequent hospitalizations, had been her life for years. Her miraculous health improvement had at that point in time been enjoyed for more than a year. Having lost approximately one hundred twenty-five pounds, she was extremely grateful. This episode occurred in the third of the four years she was my patient. She lost a total of two hundred forty pounds over the full four years.

Now let's examine that day and analyze each component further:

She called my office after a persistent compelling and foreboding feeling that she must be admitted to the hospital. Although she was persuasive and insistent, that simply was not an acceptable reason to admit. She was obviously frightened and anxious but could not discern why. Because I knew her to be honest and trustworthy, I had no doubts of her sincerity. I admitted her, thinking possibly another stroke or heart attack might be imminent.

To my knowledge, she knew nothing of my long standing friendship with Dr. Solomon or my investigative interest in the Baha'i faith, of which he was a member. She knew nothing of a celebration party planned at his house the following evening, or any of the particular details of his home. She could not have known the guest of honor would have dark hair and olive complexion. In fact, our only knowledge of the guest was that she was a midwife from England.

Her information could have only been at the most that she was a nurse coming from England. That was true, however, it turned out the guest was of Persian descent and was working in England.

There was no way in 1971 that she could have known that 36 hours later that it would be raining, but in the dream, it was and in reality, it was. She could not have known that exactly nineteen people would attend the party, but they did.

That night in the hospital, she dreamed that exact dream. She vividly saw me and Dr. Solomon open the door after the doorbell rang, and a man with a shotgun shoot both of us. She watched us drop to the floor and she woke up.

She could not return to sleep. She asked for medicine, received strong anxiety medication and returned to sleep. She dreamed the same dream with exact details. Again, she could not get back to sleep and complained of chest pain for which she received stronger medicine including a narcotic. For the third time the dream came in the exact same detail.

The next morning, she asked that I come to the hospital. I demurred as my usual visits were in the evening. She was so relentless I skipped lunch and drove to the hospital to visit her.

The other part was told in the dreamer's chapter. So many absolute unbelievable details happened to make this one of the most dramatic prophecies in modern times. Even so, the author had exclamation points. God had the dreamer experience another five dreams of untimely deaths which occurred during the following year. Each happened as dreamed. She then had a dream of her husband's death which did not happen. The dreams ceased exactly one year after the original prophetic dream. I would ask you to believe me—as much as I had to believe her—for this prophecy to save at least my and Dr. Solomon's life, and perhaps the lives of others as well.

Obviously, God has powerful ways to "show and tell." He used a full year and five accurate prophetic dreams to underscore His willful ability to produce prophetic dreams. God has said "If you seek Me you will find Me." Obviously, He is actively helpful. We must stay open to see and hear His efforts.

BEYOND THE DREAM

The dream occurred in Saint Joseph's Hospital in Augusta, Georgia. The home belonged to Dr. Solomon. Art, his first name, was a classmate of mine during medical school. By that time, he had finished his residency and was practicing locally. Adele had no way of knowing the long friendship between Dr. Solomon and myself, or of the meeting planned for that evening. The meeting which she called a party would be that very night.

Ann was one of the African American members of the Baha'i community.

The meeting was a typical "seekers" meeting. It was in fact honoring the birthday of the founder of the Baha'i faith, Bahäulláh. The guest of honor was a nurse midwife from England. There was no advanced notice that she was of Persian descent by birth. That evening a total of nineteen people came to the meeting including the guest of honor. It was raining. This was 1968 or 1969. Weather forecasts were not dependable in those days. The meeting was in every way exactly as it was dreamed—except the doorbell did not ring and a man with a shotgun did not shoot Dr. Solomon and me.

Years prior, and having personally asked God to Self-reveal, I could not ignore this unique, prophetic message so dramatically delivered to me three times—in perfect detail. It was made even more significant by the fact that the message came to me through Adele.

There is no way to comprehend this as anything but divine intervention. Adele was a patient who—through no fault of her own—had required a great deal more of me than any other patient. She had now delivered for me, and others, a significant warning that otherwise would have had no way of reaching us. This was so much more rewarding, and had such a profound outcome, that I felt it was of infinitely more value than what I had been able to offer her. I believe this to be an example of the magnification of God's Grace.

That is why earlier I said "But for Adele I would not be writing this book."

I knew that Naomi, Dr. Solomon's wife, knew more than I, so when quizzed she revealed, "We had a previous situation with our backdoor neighbors. A couple of weeks before this incident, we sponsored a similar meeting, which also was multiracial." (Recall that this was 1971, in Georgia.)

Still, I needed more details. Naomi told me that from across the courtyard the neighbor had approached her during the meeting telling her that mixed races were not tolerated in this community.

He further threatened her that if this happened again, he would be back with shotguns. Shortly thereafter they heard traffic outside. They noticed a pickup truck with men in the back of the truck visibly holding long guns. At that point everyone present was instructed to lie down on the floor where they continued their conversation until the presumed threat had passed.

The power of the spirit of that meeting had been so tangible that a few seekers had joined the faith that evening.

CONSIDERATIONS

Loosely paraphrasing Shakespeare, the same flower that we call a rose could be called by any name in any language, but it still would be a rose to us.

An explosion in any language is an explosion. Humankind has been observing explosions throughout all of our history. Consider the volcanoes which are explosions, or lightning strikes. There are eight million per day on earth—or 100 per second. Consider spontaneous explosions of silos with ignitable dust, where in hot climates many exploded spontaneously. Consider many accidental explosions of flammable material. For example; cars that wreck or planes that crash or collide, countless houses where natural gas has leaked in, combustible chemicals that accidentally become mixed and explode, munitions stockpiles that are accidentally ignited, firecracker factories that become ignited and explode, a pressure cooker where the relief valve becomes stuck and does not relieve the pressure, even a potato that is baked without a puncture in the skin will explode.

I can attest to, and have witnessed in my Psychiatric practice, that a person that is pressured beyond their capacity might well "explode." This could be demonstrated verbally or physically.

We know that an explosion releases highly compressed energy which consistently follows one law: the energy is dispersed in the path of least resistance. To my knowledge, throughout the trillions of observed or studied explosions, humankind has never found one example of a spontaneous explosion creating anything. In fact, to the contrary, it's a destructive and damaging force to anything in the path of the expanding energy. Nothing has been created, not even a penny, or anything else useful.

We therefore created a law of probability. The probability odds are ratios, every day the sun comes up for centuries on end, so we say the odds are a one-to-one (1:1) ratio that when the sun comes up, we will have a new day. If you flip a two-sided coin a thousand times, the odds are likely that about half the time either side will come up.

When we consider that out of billions or even trillions of observed explosions, not even one simple thing has ever been created, the odds of something being created from an explosion are astronomically high against that occurring. Add to that complexity, such as the richness, beauty, diversity and precise balance in the world of Nature, and you would be pressed to add enough zeros to show the odds against an explosion creating it all.

(Ex: 1:1,000,000,000,000,000,000,000,000... and so on)

No gambler would take that bet. Yet, atheists take that bet with every breath taken in a state of denial. Bertrand Russell famously answered God's questioning of his disbelief, "not enough proof sir, not enough proof." Those days are now behind us, if one is willing to see the evidence before him.

Man has learned there are ways to control the disbursement of the energy, therefore, we have fireworks with smiley faces in the sky, multicolored explosions in the sky, rifle and pistol bullets propelled in a straight trajectory—purposeful and guided by aim and force, and constant, sustained fuel explosions propelling rockets because the energy is all focused in one direction.

The intricacies of our solar system are certainly broader than our galaxy of 200 billion suns and solar systems, witness for example our universe of 200 billion galaxies. When scientists study the fine tuning of the physical laws it is statistically impossible that these connected and necessary fine-tuned laws of physics could have happened through a spontaneous explosion.

No logic can deny design by a Creator of Intelligence—even if we have not been able to detect this instrumentally. Fortunately, it does not end there, as this Intelligent Creator continues to be engaged with the ongoing happenings here on earth.

The foregoing simply references facts of the natural world around us, but they should be considered in tandem with the various examples of Creator intervention or Divine guidance in our lives that this book demonstrates. A colleague of mine, Narithookil S. Xavier, wrote a cogent book, "The Two Faces of Religion." It examines both when religion is pure it is good, and when religion is adulterated and has the potential of becoming bad.

I hope you will google N.S. Xavier. His writings on religion are vast, informative, thought provoking, inspiring and surprising. He directed and produced two outstanding documentaries. One, "The World's Most Enlightening Region" is about a city, Erumel, in the province of Kerala, where, for more than 2000 years, the people have harmonious religious groups living integrated lives. They share celebrations and activities of daily living, parks and life in general. It began with Hinduism and Judaism. Christianity was later added, and in the seventh century, Islam. It is a model for the world.

When science is wrong it misleads us. For example, I believed as did many, that the fastest thing possible was the speed of light, at 186,000 miles per second. That was for many a tremendous stumbling block for a God who should know everything going on in the universe. It takes eight minutes for the light to reach us from the sun, and yet our sun is 30,000 light years from the center of our galaxy the Milky Way which contains 200 billion suns. The oldest of these is the Methuselah star HD140283 at 13.6 billion years old, 200 light years from us. It would be more accurate to say that it was this many light years from us when its ray of light began its journey to us. When we receive authoritative information and depend on it, and yet find it to be false it is disappointing. That is why when we don't find true answers, or answers that seem to be believable, we need to keep searching until we find answers that are compatible with logic. And if that is incompatible with religion then both must be examined until they do reach a common belief.

DARK MATTER MYSTERY

Scientists are in general agreement that there is such a thing as dark energy and dark matter that we do not see, feel, touch or recognize it in our world. Its purpose and function remain a mystery for the most part; in fact, so much so that at times some scientists have even admitted "this may not be real at all." For a fact, however, it is real. Scientists are certain dark matter energy is involved in the force expanding our universe. That purpose remains a mystery.

We cannot see it, although it is there. We see through it as if it were a clear looking glass. It is to us invisible matter and energy, or clear matter and energy. Dark matter seems a misnomer to me; perhaps collectively a better name would be "The Invisible Cosmos." We only see our universe. Roughly speaking, dark energy is 70% of the total cosmic energy/matter equation. Dark matter is 25%. The precision with which matters of our universe, how it functions, why it functions the way it does, is so intricate and astounding that we feel comfortable in the calculation that all of the known matter and energy, that is which we deal with in our lives and in our universe including all building blocks, energy, all of the 200 billion stars and solar systems in our galaxy and the 200 billion galaxies in total consists of only 4.9% of the total amount of energy and matter existing in the entire cosmos. So, 95.1% is entirely dark matter/energy. There must be some purpose.

I believe God resides in and is a part of "Dark Matter." I believe God is:

Omniscient – God has an awareness/memory of everything that has happened since he came on the scene—and essentially nothing happened before he came on the scene. Not only does he know, or have memory of everything, he has perfect relational use of everything, one item to the next, engineered specifically to interact with each other, whether directly, indirectly or tangentially. He above all has known the art of extrapolation of knowledge to envision all possible scenarios. He understands the infinity of the cosmos, the ubiquitous energy and potential similar—or possible dissimilar entities—and therefore would need to be prepared for all possible challenges.

Omnipotent – Before God's first command there was no organized power, only disorganized energy. God led the way developing knowledge and using it to His advantage. To our knowledge and observation, He remains supreme and without a serious challenger.

Omnipresent – While this might sound difficult or tricky, it is possible. We are created in His image, but we are created as individual people while He is created as one Supreme Being. He is not limited to a specific area or place as we are, along with plants and other animals. Despite that, because of the vast resources at God's disposal, and particularly the way He has created us, we are in a constant ongoing exchange of information to and from His omnipresent ethereal presence. We see this more clearly in the extracorporeal vision of someone who, for example, is on an operating table, flatlines and is being administered cardiopulmonary resuscitation. Somehow, they record and recall verbal and visual activities that were happening in the room in exquisite detail. Later they verbalize and describe the details of what happened during the resuscitative efforts often in better detail than the collective staff. This is discussed in more detail elsewhere in this collection.

Omniscient – The Supreme Being has full comprehension of all knowledge. There is nothing known that He does not know and have in memory. The mechanics of this will be discussed later.

Omni-consciousness – Nothing happens without His immediate awareness and knowledge. See the following two chapters, and again, the mechanics will be discussed later.

DARK MATTER THOUGHTS

Let's consider the magnitude of the possibility. Just look at how vast and amazing the life we have awareness of is—and that's with only 4.9 percent of the energy. Even if we do not look at the material product within that 4.9 percent, if we only look at the different wavelengths of energy, even just the ones that stream around and through our body. They are practically innumerable. Radio waves, TV, x-rays, infrared, neutrons, quarks, Neutrinos, we have been able to capture many of them, but we do not materially manage all the wavelengths within our spectrum. To our credit and God's gift, we have mastered the use of much of the energy available to us.

Our mind, although a part of us, is of the dark matter spectrum. Its temporary residence is with the human to which it is assigned. Its true home is in the dark matter spectrum. It communicates regularly at God's pleasure, both in our spectrum and in the dark spectrum. One form of communication is what I call the "essence communication" and each mind interprets its full meaning in the language of the host to which it is associated. For example, if there were a dozen people sitting at a table, and each one spoke a different language, they would all immediately and simultaneously understand a message sent in God's essence language. We would each hear it in our own language, as it functions as an infallible interpreter. Think for a moment, if all messages that came to all humans came through one of our five senses, all of our brains/circuits would be in a state of constant and total overwhelm. Therefore, our mind channels to us what we are intended to receive and nothing more.

The essence of the dark matter is vast and varied, yet God has absolute resource material to design anything that even He could imagine. It could be a type of energy functioning being, or material objects strictly in those energy spectrums.

He could, for example, direct an energy force to affect positively or negatively any animal or plant. See gospel singer's miracle.

He could direct a singular force with specific DNA patterns to humans, either positively or negatively, and enhance health to the point of flourishing versus diminish it to the point of extinction.

Just as He is living in our universe and we do not see Him, He could develop a multitude of objects that we would be unable to see or discover even inside our universe.

He may develop networks through dark matter waves with speeds even greater than quantum speed. He may use these networks to be omni conscious and aware moment by moment of what is going on. He could have multiple centers of analysis and reactions based on protocols so that His will is done without His even having to be involved according to certain protocols He would have in place.

Otherwise, He may have protocols for certain situations or actions that He has funneled directly to Him as the main decider for certain situations.

He may have certain segments of energy utilized to develop innovative ideas or actions. He did that on His own—sending His own Son to be with us, to teach us and lead us.

The picture is clear. There is essentially nothing God cannot do at His pleasure with all the energy and material available to Him. His potential is limitless.

DARK MATTER USES

Several theories regarding dark energy/matter need to be developed. Here are a few.

We can believe that one of the uses of dark matter is God's "essence communication." At this time, we have not yet been able to technologically capture the communication waves that are in the dark matter spectrum. But it is a reality. Our mind, outside of our consciousness, captures images and messages, then translates them to our brain, essentially downloading knowledge we need to know, or actions we need to take.

God speaks to us through our mind, instantly translating it to our particular language and setting the information in our brain. We are generally unaware of the mechanism, only the result. People will say "God told me," which shows they are aware that the message was from another realm, even though they are unaware of exactly how it happened.

God could have one or many wavelengths of energy for example, to house and maintain a cosmic library where everything is stored in that spectrum and always available anywhere in the universe.

The Akashic Records are believed to be such a cosmic library of information, always available to all.

Another use could be an energy force that is ubiquitous strictly as a healing force that can be used anywhere at any time of His choosing.

It is possible that some of the energy could be used to effect changes in our weather.

He may be building things, new ideas, new places for angels or dreaming up new things for the spiritual beings in His kingdom that are invisible to us. Our imagination cannot begin to grasp God's imagination and plans.[15]

Our solar system is expanding, as are our galaxy and our universe. Our knowledge needs to expand. Thus far we have no definitive explanation as to why that phenomena continues. God is both infinite and finite in His deliberations and actions. Everything regarding the birth of our universe was precisely planned. What is the logical reason why everything is expanding? To understand the larger things He does, we study His smaller actions and patterns here on earth.

---

[15] John 14:2 KJV "in my Father's house are many mansions (Latin translation) or place for you (original Greek script).

When we project a logical developmental timeline for God, and compare it to earth as we know it, we see that God has utilized His knowledge, in both large and small ways, with His techniques, teaching us or by ingraining certain things in us that are particularly significant to our special needs.

As I look at the expanding universe, I am aware of the infinite space beyond our universe. In a 360 global direction there is no place where you can point that it would not be infinite. Looking outward, for example, every direction is infinite.

If space is infinite, and energy is throughout space, and God spontaneously developed and grew into the being He is, over these many years, then there is no way God would not recognize that in the infinity of space, energy and time what has happened once possibly could happen again.

If that were to happen it is logical and predictable that God would have anticipated such a happening and therefore would have had His space protected in every way, including from an invasion by a future or past super developed entity.

We can speculate that God might have shown us in part how He is doing this through the examples we see in nature. It is interesting to note that both humans and animals demonstrate a similar sort of primal need to mark off an area to claim as their own. Furthermore, animals and humans alike will endeavor to protect their space.

God may be using the force of the dark energy to push the limits of His universe as far as He wants His space to extend. He may have a plan for the dark matter to cease pushing the universe outward at a certain point. He may have a plan that when this time comes it would then be simple momentum continuing the space expansion until the momentum is spent which itself could be millions, if not trillions of years. It certainly could be seen as a way of God saying, "This space is taken. Look at its mass and expansion. Don't risk challenging Me."

Of course, it could be totally meaningless, though I have not envisioned God doing anything meaningless.

THE ANGEL R.N.

We recognize some people and some animals have something unique about them. You might recall reading about the nursing home cat that seemingly, randomly would go sit with a resident. On the third day the resident would die. We may know a great deal about behavior and intuition, but we certainly cannot fully explain how that happened consistently.

I was fortunate enough to interview an R.N. of my acquaintance with permission to share the following stories. I nicknamed her Angel. The two events I am sharing will represent more than a dozen such incidents during her career thus far. I appreciate her allowing me to bring these to print.

While I knew her at one hospital, I was aware she had worked at another hospital where similar miracles happened with her.

Typically, she would be assigned a list of patients for her shift, but on occasion fate would intrude on her day, essentially take control of her, and a miracle would happen through her, after which she would proceed on her regular rounds.

The first case shared with me was a forty-two-year-old lady with terminal advanced lung cancer. Daily, she became more concerned about this young lady facing death.

She asked God, "Do you want me to pray?" She did not get an answer until the fifth day when she clearly heard God speak to her, "Go in there, now!" Entering the room, she found a chaplain in the room berating her for her years of smoking and not taking care of her health.

Angel asked Jesus to make him be quiet. The chaplain, at that moment, turned to her and asked if she found what she was looking for. She simply answered yes.

She saw Jesus standing in the doorway. He came closer to her and the patient. She felt something like heat waves of glory coming out of her and hitting the patient.

When the waves stopped the patient commented "I see Jesus with a robe and two belts, one purple and one red." These are the colors of priests and royalty.

The patient was completely healed. The chaplain had fallen asleep through the experience.

The second incident happened at a different hospital, walking down the corridor she heard God say, "Walk in that room."

Stepping inside the room she felt something like electricity all around her. The man in bed was animated.

The patient told her, "I was praying someone would come and heal me, and you are it." A nursing student happened to be in the room and stayed throughout the event, frequently asking questions.

Angel had no idea what was wrong with the patient. There was no specific prayer request except to pray for his healing. During the prayer she saw something the size of a baseball. The baseball then shrank to the size of a golf ball and then disappeared.

At the same time she saw the golf ball disappear, the patient began jumping around in the room saying he was healed. He felt good.

Coincidentally, a CT scan was scheduled for that time. The attendants came and escorted him to the CT department as he celebrated his cure. Angel confirmed that the CT scan showed no evidence of a baseball sized pancreatic cancer that had been previously seen and diagnosed.

Angel has no answer for why these things have happened to her. She states that she sees basically the good in people and not the bad. She is a believer and knows God's power. She accepts what comes when it comes and asks no questions.

DOPPELGANGERS AND SOULS

The concept of doppelgängers is an interesting one to investigate further. Can these two double goers actually be the one and the same person, mind and brain, together and separate?

Scientists now claim to show the brain functioning at quantum speed. This is a misinterpretation of fact. They speak of the human brain—which is flesh and blood—and which supports speeds of only approximately the speed of sound. However, they would be correct to say that the human mind can function at quantum speed. This is possible because the mind of humans, (and evidently animals as well), is dually functional.

The mind usually resides in the brain, its natural host. We now have scientific evidence that it is extracorporeal-digital, and may at times exit the body intact and return at will. It is capable of quantum interactions while residing in the brain. Any calculation or information it receives could be simultaneously downloaded to the human brain. An investigator of the human brain would not be aware of the extracorporeal, invisible input.

The word Doppelgänger is from German lore because of the many tales of persons who would see or know things, some even at great distances, that were impossible to naturally know. The most famous American doppelgänger was Edgar Cayce, who was so phenomenal that a foundation to study his life and work continues to exist even today. He had modest education and no medical training or background. With a piece of a personal belonging from a sick person, even coast to coast, while his body remained in a trance his mind would travel to that person, find them and start by saying, "we have the body." He would then describe the person, the condition, the illness, the diagnosis and the proper treatment for the person. This became such a phenomenon that the medical profession forced an investigation. He was found correct in 97% of his readings. Stories like these from subjects far and wide are well documented for your further study. One cannot rationally deny, nor by instruments prove how these occurrences happen. You can only by mathematics examine the odds, which leads to wonder at the chances of this happening in the natural world.

Although the extracorporeal mind of Cayce would be obviously at the bedside and doing the work the corporal brain and body of Cayce would speak as if it were there at the scene describing everything that happened. This transmission of information at these distances gives us an immeasurable glimpse into just one potential use of dark matter.

Most of us, especially those of us in the medical field, have heard about or know of cases where it is said that "people die on the operating table." More realistically, I believe that we should be saying "life is suspended" and when the patient is revived and awake, they are then able to, in great detail, describe everything that was said and the activity that went on during the life suspension. This includes who said what, and who did what, as if the person was visibly viewing and recording everything from a perch above the scene. These incidents have been verified and recorded on multiple occasions. In fact, just today I heard a lady tell of this exact sequence happening to her sister.

If it were only verbal, we might not be so surprised because we know that patients under anesthesia are hearing and subconsciously recording what is being said. But, to have the words of every person in the room described as to which one said what, along with the visual description of everything that happened is not explainable except with an extracorporeal mind watching, or in other words the patients mind has left the body for a time, yet was still "present" close by, as an observer. Upon returning to the brain, the information shared was brought back in as well.

The case of a pilot participating in an air show is eerily similar. His plane crashed on the runway with spectators on the edge of the field. His body was pulled from the plane, apparently dead and laying on the runway edge as attendants dealt with the plane.

After several minutes, the pilot got up and walked directly to a lady spectator at the edge of the field. He said words to this effect to her, "Ma'am, I want you to know that I am not a stupid pilot. The crash was not my fault. It was caused by a mechanical error." (It was later verified that he was correct.) According to the report that I read, the woman fainted. Those around her confirmed that she had made those statements to them, but how could he have known. His simple answer was, "I was here and heard her." I apologize for not being able to identify or remember the source of this story except that I remember reading the report some several years back.

For an example closer to home, I will share two incidents which I personally experienced.

The first is a short exchange that perhaps could happen to anyone. To me it only happened once and, though brief, should open one's eyes to the fact that there is so much more to human beings than our flesh and blood. (The previous incident involving the pilot also demonstrates this powerfully.)

The second one is truly one of the most profound and compelling experiences of my entire life. It must be the lynchpin key proof that in fact the doppelgänger mind is the eternal soul. It is the introductory miracle at the beginning of the book – The Multi-Purposed Miracle.

A DOPPELGANGER MIRACLE

A sixty-two-year-old lady was in my office for a routine visit. In the course of the conversation, she mentioned how blessed she felt to feel so good. I commented that I could certainly understand remembering how sick she had been when I met her.

She reminded me that she had bled so much from the huge ulcer that she almost died. In fact, she said, "I did die, but only briefly." She continued, "I saw my daughter and son-in-law. They were terribly upset. He was in front of her, holding her shoulders and trying to console her."

She explained that her heart had stopped. She had not realized it because she was so sick, she was unconscious. The attending was able to resuscitate her to an eventual recovery. I saw her a few times for some issues, and she continued to follow me at six-month intervals.

She mentioned, "You know, the world is in such a mess right now, I believe it may be the end times. But I am truly at peace now since that episode. I felt so totally at peace. I know that I left my body while they were trying to revive me. My daughter and son-in-law were in a different part of the building, yet I was with them and not afraid. That is all I remember until I woke up from the ordeal. Later I explained to them that I saw them and what I saw. They were astounded and stunned that I had described exactly where they were, their position and what I described was correct. It gives me peace to know that my soul will be free and happy when my time comes."

Although her story was a simple brief glimpse of the doppelgänger effect it was nevertheless extremely profound to her and meaningful. She had not needed it but considered it a miraculous gift to feel and know the peace of the afterlife.

This story was an unsolicited gift to me, and I give to you, the reader. It is one more example of our extracorporeal twin that apparently is immortal and free.

FINDING GOD

No human except Christ could give us a full description of God's makeup. No one should claim that ability—neither through science or faith.

We know bits and pieces by His words and actions. And we know some things by deduction, speculation, or through extrapolation.

These thoughts have come to me not by my meditation or search, but by appearances from time to time in my mind. I have assumed them to be knowledge periodically provided by God in answer to my prayer: "God, if it can be in Thy will, I would like to know more about You, our purpose and the afterlife."[16] I have wanted to share this knowledge, but at the same time I have been unsure about whether or not I should. Meanwhile thoughts continue to pop into my mind.

I believe that not only does God know what I am writing, but that He is okay with it, and is aiding me by way of these visions,

---

[16] Matthew 7:7 "Ask, and it will be given you, seek and ye shall find; knock, and it will be opened unto you." KJV

thoughts, and insights. He understood my doubt whether or not to share them. In what was to me an epiphany I honestly believe that God allowed me to know that it was okay to finish this writing and publish it.

Therefore I have decided that, in a later chapter, I will share the epiphany that formally launched this collection toward publication.

PRIMORDIAL ENERGY / MATTER

Primordial energy—as considered before there was a universe—reveals that there was nothing but space and energy. Possible exceptions could have been tangles of energy forming pebble like objects, but made of energy rather than mass. Then God arrived and eventually conceived, planned, created and nurtured the universe.

How did God create all the mass required for the universe? He learned how to manipulate energy from wave to mass. It would be unreasonable to even speculate how many years or light years God might have used to learn everything he obviously knew when He created the universe.

There are various theories on how he specifically brought this universe into being. One of them is a collapse of what He had before, and then exploding it in the 'big bang' with all the laws incorporated into it. As huge and expansive as the universe is, ultimately it is an island sitting somewhere in infinite space.

We know that energy was in the cosmic space before God created the universe. We do not know if it contained all the wave forms that the universe has (with our five senses detecting only 4.9% and the remaining 95.1% tangible only to God, and to whatever support system He has developed.)

Some far-out thinkers believe we will harness enough knowledge and control of energy that we will be able to either challenge or sidestep God in controlling the universe. I can imagine God laughing at that. After all, He controls 14.5 times the energy mass force that is in the universe total. Furthermore, He has control of the cosmic space and energy outside the universe. That is where His initial home base has been and that apparently is an infinite field of energy in the infinite space.

Many questions are unanswered regarding the primal energy field. For example, it may be all one energy form. God may have tweaked it to give it various wavelengths, because the more wavelengths available the more utilization we can eke out of it. We know for example that He has given us wavelengths in such ways that we have been able to use in numerous applications, from x-rays to heat, sound waves, light, condensed such as in lasers, flashlights. We can assume that He is able to do that with the rest of the wavelengths in the 95.1% which would make potential uses practically uncountable. Remember, however, He is not limited to this universe. He was "born" and matured and self-educated in cosmic space. There was no universe until He created it.

The marvels of what we will discover going forward or after we transition to the spiritual world are certainly exciting to contemplate.

EYES OF THE BEHOLDER

I'm always fascinated when especially meaningful events happen. Sometimes these happen in our spiritual life as well as in our secular life. The event I am about to share with you is one such event.

In early June 2019 I chanced to see a Christian friend familiar with my efforts working on this book. We engaged in brief catch up conversation. My friend then gave me some words of encouragement for the book. Then, once again related that he remained skeptical regarding my interpretation of events in my life regarding clouds. One of the things I like about his Christianity is his absolute honesty. I had given him copies of a few chapters which mentioned two or three of the things I consider miracles in the clouds.

What happened next finally dawned on me the following Monday. It is one of the things I absolutely love about God. Saturday two days prior had been the day of our encounter. The weather forecast had been terrible for the weekend, forecasting three inches of rain in fact. Saturday late afternoon when I arrived home my wife wanted to show me something. She took me to the master bathroom and pointed to the skylight. I was surprised to see that the dome part of the skylight had been shattered. The glass bottom of the skylight remained intact. I did not appear to have water on it. No water had leaked into the master bathroom. We knew that more rain was coming. This means that the rain had started during the day of our conversation. I had spent the day at a complex some ten miles from home. There had been some rain, but not much, but more wind. A limb had broken off in the wind and blown over to fall on our skylight breaking the dome, but not the glass of the ceiling.

The rain had been fierce and heavy at the center where I had been, and widespread. Roughly for a couple of miles around our home the rain was light, but the wind was strong.

This set-in motion, as you might imagine, a concerted effort to protect our master bathroom knowing that a heavy rain forecast remaining would, by weight alone, break the glass that had so far protected us from shattered glass and water on and in our carpet.

We were facing a huge problem. We called our last contractor. He stated he might be able to come help after church Sunday morning. Sunday morning, he ran into someone at church that said they would do it. He turned it over to them. We called the roofing company Saturday but only got an answering machine and no return call. The person our contractor spoke with was indeed willing, but we knew several strong reasons that person should not attempt the task we needed. We were back to zero.

Sunday afternoon I bought a tarp in case we could find someone; we knew the rain was coming. Neither of us could find anyone that could come and place the tarp over the skylight.

Sunday, I had an afternoon appointment at the office which lasted until 4:30 PM. It was not raining. Approximately forty-five minutes later the rain was coming down more fiercely and harder than I had seen in a long time. Walking to my car, in the sloping parking lot, was through water more than a half inch deep. The rain required a fast windshield wiper speed and still it was a difficult drive. The drive home is approximately seven miles. The rain was extremely heavy for the first three and half miles. The rain began slacking off gently so that when I arrived home the rain had barely been slightly more than a sprinkle, not one drop had come into the skylight or our house.

This was one of the heaviest rains predicted for our area. It happened as predicted, but miraculously skirted our house.

Before 8:00 AM Monday a call came resulting in a tarp being placed to protect us. I was happy to get this repair done. I consider it a constant reminder to me of God's omni-consciousness and omnipresence, and His awareness of what we need in order to know His power, intimacy, reality, and willingness to intervene. The odds against that being purely coincidental are simply too much. This all happened within a few hours of my friend expressing his skepticism. I believe that the way it happened cannot be anything but a personal message to me, belying that skepticism.

This is my friend's personal miracle, but I think it was intimate and meaningful for both of us.

COMPLEX INTERCONNECTEDNESS

One scientist whose name escapes me, mesmerized by the power of computers, announced he would start a new religion using a new supercomputer as the go-to answer for everything. Although humans have progressed dramatically in computer development, even developing artificial intelligence, we cannot possibly reach the level of a creator who is proven to be literally "light years" ahead of us with His thoughts and His accomplishments.

Further thoughts on dark matter:

Scientists have figured out that what is named dark matter represents 95.1 percent of the total of what they are calling "matter," which is all invisible energy and matter combined. A proper name for this should be "non-visible energy matter." Matter is what we describe as sensibly detectable by touch, sight, taste, smell, or sound: everything that has been transformed from energy into matter. This meaning ignores dark matter which exists, but eerily—despite its mass—remains invisible to us.

When we look out into the sky especially with our sensitive and powerful detection devices, we see billions upon billions of stars. Each one is a mega conversion of energy into mass and then we realize that space goes forever and there are stars throughout our vast universe. We must be awed at the possibilities of development with so much invisible energy everywhere in space.

We realize there is an interface that is not visible, but is usable, just as is visible energy. Light and heat rays for example, and particularly radio and TV wavelength spectrums. These do not register in any of our senses, yet we capture and use them. We control this energy in unique and powerful ways, no matter the intricacy of a picture. For example, imagine how infinite color changes can be speeding through the air—even constantly flowing through our body—going long distances through many humans before reaching an impenetrable solid material or a receptor that can completely perfectly reproduce the image that came from anywhere, at light speed, 186,000 miles per second.

For a singular God to be able to keep up with and interact with everything at once, there must be the most advanced supercomputing mechanism imaginable. This is the essence of God's ability. There must be a communication mechanism that can connect with, communicate with, and interact with, all instantaneously. Thus far that has not been elucidated. Carl Gustav Jung, son of a theologian, and a 19-year younger contemporary of Freud, and Freud himself, both being psychoanalysts, mentioned a "collective consciousness." This may be the closest recorded reach toward understanding this theory.

As an aside, some computers send messages to other computers to receive information but do not receive information back from those end products. Other computers send and receive information from each other. This is a mini replication of God's computing capacity. Examples are lower life form, amoeba, birds, dinosaurs, plants, flowers, trees, and food are receptacles for this computer technology and the intricacies of these living receptacles can be programmed from a distance, but they have no consciously interactive interchange ability. These life forms may be close to having a collective consciousness, as plants and animals do seem at times to sense what is hospitable or hostile. The next chapter displays this reality to those who are naïve to this.

Therein lies the truth—and the biggest distinction—with the human being created in God's image. Humans house a minicomputer that is interactive with an information exchange mechanism of God. That ability is through the mind, not the brain. As we have seen elsewhere, the mind resides mostly with the brain but can separate from the human body intact and return intact. The mind can receive and send information at quantum speed at any time including when the mind is seated in the brain. Whatever it sends and receives is instantly available to the brain at the discretion of the mind. The brain itself cannot replicate that ability, neither can the lower animals because flesh and blood function at more modest speeds, not even close to the speed of light, let alone quantum, but specifically only around 275 mph. Yet, science has clearly demonstrated the brain can appear to calculate at the speed of quantum. While it cannot, the mind can, and the brain therefore appears to be doing the function because we are instrumentally unable to separate the functions that are performed between the mind and the brain.

It is interesting to note, while our chemical neuronal system is relatively quick, our body is a good conductor of electricity at nearly the speed of light. When electrodes are placed on each temple for electroconvulsive treatments with a good conductive medium, usually a gel, they are held tight to the temple for the electrical conduction. The electric current touches the skin first, then the subcutaneous tissue, then the skull, then the dura, then the brain. When the current touches the skin, a message is put in route to the brain traveling at 275 mph. However, the electrical current is traveling at nearly 186,000 mph. The brain is instantly asleep and feels absolutely nothing, even the convulsion or any pain, and no memory of the procedure is recorded. Incidentally, the same principle holds true for the electric chair.

To illustrate the concept, let's picture ourselves somewhere, anywhere, on the deck of a ship in the middle of the ocean, in our backyard in our lounge chair, inside our home or office, any spot! Think of what is streaming past you and through you at this spot. Light waves are streaming around you and on to you and stopping at you, being absorbed by you. Radio waves, TV, x-ray pass right through you. However, when they encounter a receptacle such as a radio or TV they are captured, converted to usable matter such as pictures or sounds, or programming a computer or other device.

For example, I asked the audiologist to reprogram my hearing aids. She did not ask me to remove them. She simply accessed her computer keyboard and reprogrammed the computer simultaneously reprogramming the hearing aids I was wearing. The radio waves coming to my hearing aids left the computer in a 360 spherical pattern going through my entire body without being stopped except by the hearing aids which stopped them. They went through the audiologist's body. They went as far as the provided energy would push them.

The energy of the sun pushes light rays to warm us and our earth. Light waves from distant stars, even light years away, reach us having traveled through space without being stopped or interrupted. Yet, when we see it, it is stopped even though other wavelengths from that same star might pass right through us.

There might be a dozen TV stations and 100 radio stations in our city. Each of them beams out information verbally and/or visually. These waves are ubiquitous around us. When they go to a TV or radio, the only ones that stop are the ones tuned to the frequency selected on a particular device. Others continue their merry way, absent the presence of non-penetrable solids.

Our mind is equally fine tuned. The wavelengths coming to our mind knows our "address," so to speak, just as the radio waves "know the address" of the frequency they are attuned to.

We must also point out that if you consider the rays of the sun falling on different textures or colors, varying degrees of heat will be absorbed, according to both the texture and the color. Furthermore, radio waves are captured on receptacles at the same frequency and even then, there are different grades of receptacles. For example, the older TV's, the newer TV's and the high-definition TV's. Our brains are no different. Some brains are finer tuned than others. Some will be able to receive messages that others could not. Two examples of this include one from my experience recounted in other chapters. The most dramatic demonstration of this was Jesus talking with the "woman at the well." His mind was the perfect replica of God and therefore could call upon the full repertoire of cosmic knowledge and likewise cosmic power. That is, to be able to tell this woman her history in detail or to change water into wine at a wedding.

See chapter: A Different Look - A Parable.

GOD'S POWERFUL FORESHADOWING

The title is not meant to be flippant. God's enormity and existence cannot be seen with our eyes or heard with our ears. In my estimation it is God's genius to be living in the dark matter portion of the energy spectrum. He has total comprehension and vision of what we do but our vision of what He does is not so complete.

God does say "you will seek Me and find Me when you search for Me with all your heart."[17] He shows Himself, through His powers, in many ways. He foreshadows events, preparing us to understand future meaningful occurrences. He shows us one miracle in order to help us understand another.

Two Biblical examples are particularly interesting in this manner. The first was God speaking to Sarah, telling her that she and Abraham were going to have a son. Yet, she was in her nineties. She laughed. She said she was past her time. She was not of

---

[17] Jeremiah 29:13

childbearing age then, and had not been for many years. God was undaunted and explained the son she would have. Of course, as we know, Sarah had this son as was foretold.

Think of the power of God in having this happen. Her bio-physiological body had to be rejuvenated. This is an impossible task for humans, but God showed it to be possible for Him. It was not the mere fact of having a sperm implanted and her becoming pregnant. Her entire reproductive system had to be restarted and maintained and at her age, she had to be able to withstand the pregnancy, the child's infancy, and to survive all that was required to raise her son, Isaac. Several centuries later the angels told Mary that she would have a son and explained his name and his mission. Yet, she was a virgin. God's son was born of a virgin's womb.

It is important to note that in both instances these occurrences are beyond science, and yet somehow were still foretold before the incidents happened. The prophecy must have seemed absurd at the time, making their fulfillment all the more miraculous.

A similar happening was when Abraham was following God's command to take Isaac up the mountain to make a sacrifice. At the sacrificial site Isaac asked his father where the sacrificial lamb was. Abraham's answer to Isaac was "God will provide." That is exactly what happened and at the last and critical moment, the lamb tangled in the bushes, bleated, signaling God's provision. The takeaway here is not the fact that He was going to have Isaac sacrifice the son but Abraham was such a believer and follower He trusted God completely to provide what he needed.

The story leads all of us into a terrible feeling that God was going to have a loving father sacrifice his own, and only son. We were horrified reading that story, only to be relieved at the end. Then we came to realize Christ was sacrificed by His Father, God. We were able to recall that scene and realize a measure of what God was feeling because He had foreshadowed this event centuries before.

It is awesome to think that God could at the time of Abraham predict that He would be sacrificing His son for us, thousands of years later. See the next paragraph and "Stealth Scientist" chapter. This shows that when you have all the information and knowledge there is, you have a tremendous advantage in predicting the future. So it is that He was able to accurately predict the future. He lets us know free will by showing that at times He answers our prayers positively when it would not appear that this was planned centuries before. One simple example from my life was the rain stopping as I was waving a magic wand asking the clouds to stop raining.

More particularly is the scientific message given to us in the great pyramid of Giza, containing many accurate scientific statements and many prophetic predictions that were "pinpoint" accurate, including: the exact day Christ was born and the day He was crucified, the exact day World War One was declared and the exact day its armistice was declared, and much, much more.

Humans alone could not have predicted or orchestrated those events. But an omniscient, omnipotent God could infuse that knowledge in the human mind that would educate the human brain, that would then translate it to the building of such an instrument of knowledge as the pyramid, being four thousand five hundred years before Christ. Also remember the Hyksos builders of the pyramids were descendants of survivors of the destruction of their civilization "mu" five thousand years earlier. Those many generations of descendants were able to keep the scientific knowledge intact without adulteration for five thousand years.

Regarding prophecy, Genesis lays out the exact chronological order in which the universe, particularly our solar system and our earth came into being. Chronically scientists agree one hundred percent. Carbon dating and other studies point to the "seven days" being metaphorical.

All this leads to revelations which seems to indicate that we are at the end time, but other Scripture tells us we do not know the end time. So, we live and work and hopefully grow in spiritual stature irrespective of when the end time will occur.

IN HIS IMAGE

How should we go about thinking of this? I believe in several ways.

First, developmental. God was born of the only thing available which was energy, albeit a tangle of energy. He essentially went through similar stages as humans, that is,

> infancy,
> reacting to His environment,
> childhood,
> exploratory learning,
> adolescence,
> streamlining for efficiency[18] - shedding excess framework,
> becoming Self-aware as an individual with beginnings of ambition,
> teenage years,
> learning more about self,
> maturing, and
> becoming a 'Scientist Extraordinaire' - a creative intelligence beyond comparison.

---

[18] Our teenage brain sheds excess – in the way – brain tissue

I conceptualize that at this point there was recognition of the haphazardness of the mass accumulation being disorganized, and having the development of thought processes being accidental. He in the teenage years chose to streamline His being by carving away the extraneous collective tangles as they would be cumbersome and useless.

The human brain goes through a shedding process during the teenage years whereby useless brain tissue is shed and streamlining is allowed.

Adulthood would look something like having been self-taught, now much more educated and experimentally learning such things as responsibility, organization, storage and retrieval techniques. I speculate that God's learning was by logic and by observation. Furthermore, I would speculate that since He is a person of pure organized energy which in no way would have the concept of anything outside direct logic. There was no emotion and that God observed emotion in the lower animals including anger, jealousy, rage, sadness and He then incorporated these into His own Being and perhaps that of His heavenly host. Further, He maintained His power and control so that He would be above His own laws. He, however, would remain supreme and not put others above the law.

I believe God was eagerly awaiting the creation of the humans. He had already created a genetic template but to let it play out might take many centuries longer.

When Ford invented the Model A, he did not have to go through models B through S to develop the Model T. He simply jumped that gap, so we do not know what models C, D, etc. would have looked like or been like. Those are "missing links". Likewise, there are our own missing links. In the next chapter, you will read about a chemistry professor who gave me an A for "jumping a link." God saw what He wanted and commanded it to happen. I believe He saw our image centuries before he brought us to fruition. Once again, we see God's power through what we call the "missing link."

I believe God truly gives us the opportunity to believe and to join Him in an afterlife. He did give us free will and generally appears not to interfere. In some cases, He makes non-belief nearly impossible. He does say, if you seek Me with all your heart you will find Me. That seems fair enough. He went much further, sending His son to be a human representation for Him. The fact that He allowed His son to live on earth in our image leads me to believe that we are the image He has chosen.

I am convinced by words of the Bible and by His infinite grace for us that we truly are made in His image.

There is no question that we should strive harder to function in His image.

SCIENCE HELPING SCIENCE

In my first year of college the staff and the students had a collective open forum discussion once per week. Students or staff could present questions or topics.

One student raised his hand in context to a question about creation. He considered the fact that we did not know the length of the days when creation happened and when the human race was in its early formative years. He wondered if perhaps the days had been much longer.

I raised my hand and was called upon. I noted that according to the encyclopedia Britannica the earth's rotation slows at a terribly slow rate, at the time I remembered the exact amount. The current estimate is one-millionth of a second per day.

A chemistry professor approved my answer with comments that factual answers are preferred and useful but answers that are simply conjecture are not useful.

The interesting thing here is this professor was an extremely hard professor—consistently giving a large portion of his classes low grades, D's or even failing, while the others in his classes rarely received high grades.

When I began his chemistry class, the professor remembered me. Fortunately, I had a reasonable penchant for chemistry, so we hit it off quite well.

During the class he gave us a chemistry problem to work over a weekend. It was a difficult problem, but I had finally solved it. On Monday when he asked if anyone had worked on the problem and arrived at a solution, I was the only one to raise my hand. He was shocked that anyone had been able to do it. He asked me to put the formula on the board. I wrote out the entire formula. He did not interrupt me but at the end he said that this was amazing. I got the right answer. However, he pointed to one of the steps in the chemistry formula that would not work. The progress simply would not happen the way I had it. He was still amazed that I got the right answer. (I had somehow effectively jumped a step, or created a situation where there were missing links, and yet still arrived at the correct answer.) I also got an A in his course. Furthermore, I got a one-on-one serious conversation with him, asking me to reconsider my plan to go into medicine and to pursue industrial chemistry. That certainly was one of the best compliments of my college career.

The main takeaway is that it gave me an a deeply personal enlightenment and insight which you will read about in other chapters.

ADAM AND EVE

One of the longest running discussions in the world is this: did homosapiens evolve from the amoeba, or were they created by God? The answer is surprising. Evolution progressed toward homosapiens, but it ceased at its current highest point centuries ago and has not evolved further. It is as if there has been a freeze put on evolution.

Adam and Eve are a separate creation far removed from the last evolutionary iteration demonstrating a genetic chasm that would be impossible to evolve in one step. Each genetic iteration might take many centuries. Each gene for example might have few or many intermediary changes throughout evolution. Therefore, the term "missing link" is a misnomer, for there would be multiple iterations of change required to traverse such a tremendous gap. At the very least, the term should be plural: "missing links."

The one alternative, homosapiens, specifically Adam and Eve, were created as a couple from a similar template, but not by incremental changes. They were created with the same ingredients, but a different outcome.

Experiences in our lives help us find these insights. My getting the right answer by skipping a step in a chemistry assignment and being rewarded by it, helped this insight to evolve. When Ford Motor Company built its first luxury Lincoln Continental, though it had some features of the earlier Ford and similarities of other automobiles, the Continental was a distinctly new brand.

Why and how the homosapiens image came about of course only God knows, but he allows us to speculate. He tells us through scripture that we were created in His image. We are allowed to infer that as an end result of perhaps many centuries more, (if there were not a freeze on evolution,) possibly we could have come to the same conclusion, homosapiens, but we can never know.

It appears to me that God knew exactly the end product of each step. He liked this image as it appealed to the way he viewed Himself. He took the image as His favorite. He planned the human species to be far more like Him than any other creature. He simply interrupted evolution, and created the homosapiens entity as planned.

If some doubt His creating a full person, we need to look no further than the long history of angels appearing in full form and then disappearing. In fact in this book I relay five firsthand telling of such miracles. I consider it an extreme honor and privilege to have been that close to those miracles, not to mention the ones that happened to me. In each of the five miracles I am speaking of, an angel appeared in actual physical human form, performed some miracle, and as quickly as they appeared from nowhere, they disappeared into nowhere. Miracles like this have been recorded throughout history. No doubt, as we become more attuned to knowing they happen and being ready to receive them, and notice them, we will be blessed more with their presence.

Some might ask if these occurrences were real. Recall the miracle where the angel appeared to save the little boy from falling over the escalator rail. She appeared from nowhere, in full physical form, grabbed the falling child, returned him to safety, and they disappeared. This was witnessed by several people. We see in this instance, God in His able assistance was perfectly capable of immediately producing an angle as solid and perfect as any human, who was physically present when needed, but then was instantly no longer there.

The lesson of course is that if He can make people who last only a few moments, but who are fully human and functioning, then certainly He can make people and allow them to continue living a long life on the earth. So, yes, God created Adam and Eve. He used some of the same formulas, only tweaked, that He had used before, but toward the image He was always building for. He just cut out several steps (jumped the missing links) and created us to be in His image.

Our reproductive stages of infancy to adulthood are matched identically to God's own development. Even though Self-taught, He has remembered His stages and has programmed them into our development listed elsewhere. We have deduced our developmental stages parallel to His. After you have read each of the stories in the "Angels Among Us" chapter, it will be easier to see these as a coherent, creative plan.

In considering the tremendous gap between chimpanzees and humans consider this: it is as wide as the wooden stick tool they use to rake grub from tree crevices to a modern Boeing 747. Left to the process of natural evolution, a jump that profound would take unimaginable centuries.

GOD SKYWRITING

You may be tempted to think that what I am about to share with you was a hallucination. It was not.

God foreshadowed this quite well. Remember His stopping the rain instantly on request when I was in college, and it being witnessed by a fellow student? Remember Him on two separate occasions causing a selected cloud to disappear, upon my request?

My personal experience with hallucinations came in a postoperative situation when given pain medication. There is a huge difference between reality and induced hallucinations. The hallucinations were medically induced and accompanied by a weird feeling of knowing the unreality.

This miracle was well scripted and masterfully delivered and something I could not have dreamed could happen. It was in fact a message which took days to figure out. It was divinely conceived and scripted, produced dramatically, delivered personally and intimately, and then signed by God.

AN EPIPHANY

In the early 60's I asked God if He were truly there, to please show me. I was about to raise children in the Church, and I wanted to know if it was the right path. He did, many times, and I am happy to share them in stories revealed throughout the book.

Years later in 2011, as a solid believer, I asked God, if it could be in His Will, to please allow me to know more of our purpose, how we came to be, and how that relates to matters in the afterlife. What is life after death? Again, I found that God has not hidden Himself, but shows us plainly who He is, and where He is, and reveals Himself nicely through science as well as faith.

Having mentioned to friends and colleagues my consideration to put this in a book, the overwhelming majority said they wished I would and look forward to reading it. A few, very few, simply said "good luck." Even fewer simply looked at me blankly without a word. My real question was, would God be okay with me revealing these insights of Him and eternity in published book form.

I say sincerely, the following came to me exactly as I have written it. If this event was not the validation I sought, at least I have perceived it as permission for publication.

On the afternoon of June 25, 2017 my wife and I were in route to the beach for some much-anticipated family time, which for most of the family means the beach, pool, fun with family and games. For me, it was less of that and more of reading, writing and editing previous notes.

Relaxed in the passenger seat as my wife drove, I looked ahead and, seeing enormous clouds, I commented that I had never seen clouds like those. The entire sky was filled with clouds: large, small, medium, dark and white. Directly in front of us was one of the largest clouds. The bottom of this huge cloud was entirely smooth and flat across the horizon. Underneath there were three lines of white clouds, one on top of the other, consisting of three rows of black boxed spaces. Inside each block was a distinct 'letter' by a white cloud, but in a script unfamiliar to me. Three rows that basically ran from the left of the windshield to the right of the windshield. I marveled that they looked much like a foreign language, or as I thought an ancient language might look. Though amazed, at the time I suspected nothing more.

I mentioned the unusual cloud formations to my wife, who took notice.

I looked out the right window and saw—amid dark clouds—an exceptionally large circular white cloud so brilliant it was radiant. It was twice the size of the sun. When I looked back to the front, the clouds and 'letters' remained the same.

Returning my gaze to the bright cloud I saw it now framed a cross. On closer inspection it was a five-pointed star, with a short top ray, a long, downward ray to its right (my left), and the ray on the left upper side was stretched out straight and shorter than the right. Then, from the bottom there were two medium long rays. After viewing this for some time I turned back to the front. By then I was intrigued.

On the lines of letters in front of me I saw on the bottom line, a little to the right of center, three distinct letters: S I S. Further to the right and following several indistinguishable letters was the number "8," lying on its side. At the time of recording this, on June 26, the following day, I had no knowledge what any of this meant. I did however wonder, should the letters be read left to right or right to left.

When I turned back to the star, I was shocked. The center was no longer a bright cloud but had morphed into black and white clouds. I was plainly looking at a face on a head in the position of hanging down to the right. The face, I must tell you, even if I were an accomplished artist, I could not find a possible way to paint, or write words to adequately describe the sorrow and agony I witnessed. The pain appeared to be in the present time. Suffice it to say, empathy rushed over me, and touches me still when I think of it or write of this encounter. It felt as if I was witnessing Christ on the cross. Even recording the event a day later, I continued to see the face vividly and feel that sorrow.

The letters and the lines persisted, while their meaning continued to allude and puzzle me. "S-I-S"… I have a sister, but no attachable meaning surfaced regarding her. Nothing else significant came to mind. I was puzzled. I tried to find other recognizable letters, but there were none other than the clear "S-I-S,"

I then reluctantly returned my gaze back to the face. This time what I saw shook me to the core, but in such a wonderful way, leaving an EPIPHANY of comfort. There was a large perfectly round hole in the sky, where the face had been, now solid blue with three white clouds, forming three distinct legible letters, all capitals, G O D.

Astounded, I mentioned the last scene to my wife. She commented, "God is trying to talk to you today!"

As I mentioned, this happened on June 25th, 2017, and was recorded in my writings June 26, the following day. On June 27th, reading over what I had written, I was jolted in a way I had not previously felt. I had been struggling for at least three years over what I should title this book. I had come up with at least a dozen possibilities, but not one satisfied me. Re-reading my notes brought back the visual of the "S-I-S", and without any cognitive effort on my part, I suddenly knew that reading left to right meant "Scripture Is Science" and so reading right to left would be "Science Is Scripture." Puzzle answer revealed!

I have not thought of three other words that could be so meaningful, and read well from left to right and right to left. I believe I was Divinely given the title of this book, because that is exactly its true essence and message. On relating this story to my daughter Janine, she instantly reminded me that the "8" on its side symbolizes infinity.

God demonstrated His power to intervene at will through communicating with me. I understood Him to be the original and best author, and the original and unequaled scientist. He identified his compassion for us. He left the clear message that His future is infinite. The many blank letters I believe was His way of showing us that He is for all of us. The future is infinite, and there for all who choose to be with Him and share in His promise.

This peaceful message from God, using an array of clouds, miracles and interventions demonstrates, at least for me:

1) God is here with us, and for us. His interventions can be so close, personally directed and intended for one or many.

2) His interventions can be quiet and almost imperceptible, but they are often dramatic and unmistakable.

3) Even though we have vast literature in history and philosophy, real meaning can be found in the scripture and His works, which science is now unraveling for us.

4) I believe Christ's human experience was important in His ability to feel with us and for us, but also to continue to feel the agony of our continued sinfulness.

5) His omniscience and omnipotence.

6) The research and study of this topic reveals the dedication and brilliance of the community of our scientists as they continually discover the laws for our universe set in motion at the moment of creation 13.82 billion years ago. Noteworthy, they work! As humans we yet have not produced one physical law.

7) Now we know and understand more than enough to see the Father, Son and Holy Spirit interacting relentlessly on our behalf. Much as He pursued Jonah, He pursues us still today.

8) It is a painful thought to realize, just as an identical twin in the Midwest woke up believing she was on fire at the time her twin was being buried alive in the Canary Islands where the plane she was on crashed and burned, God was feeling every pain His Son was feeling. They did that for us, for our salvation. Believers should do everything in our individual and collective power to make sure everything

possible is done to provide every single person enough knowledge to believe and accept this gift from God of love, grace and compassion to an eternity in heaven.

9) If God has a thought and wishes it to happen, that becomes his volition and it happens exactly when and how He wishes, as easily as we turn on a light.

GOD'S ARRIVAL

The following theory and others mentioned throughout this book did not come to me in my contemplations of the science/religion conundrum. They came as bits and pieces in random situations and places. They are presented as plausible and reasonable, not disprovable or provable. They are stronger than a hypothesis[19] as they are the analysis of a set of facts as they relate to each other. In our case miracles play an important role.

For example, the following theory popped into my head followed by a set of facts that together strongly supports this theory. Over time more and more thoughts came to bear, embellishing its credibility. An example is thinking of God as infinite. We project His infinity going forward. Looking back, however, it seems totally impossible that God could have been here before space and time. Therefore, logically He had to develop in time and space, meaning He had a finite beginning, although an apparent infinity going forward from his infant inception. Somewhere in the past, probably trillions of years ago, there was no such thing as a God.

---

[19] An assumption, awaiting testing.

We know that God is at least as old as our universe 13.82 billion years old which is nothing compared to the trillions of years prior. How many years would it take to self-discover and invent all the knowledge required to create the universe? We have no way to calculate. We can with confidence theorize there was indeed a finite moment when tangles of energy became aware of Self and that was the birth of God.

Colloquially when we speak about God having always been, we are talking about our universe. Surely people do not think that God was here at the beginning of time and space. If that were the case, He would have existed without a place to be. If time were not yet, there would be nothing. He would have to invent and create space and energy but His existence would be impossible.

As an accomplished magician having performed on stage and in private, and over time attaining recognition to the order of Merlin, I can attest to what every thinking person should logically know: there is no such thing as "poof the magic dragon appears." Not even a rabbit out of a hat.

A magician's performance is a disciplined practiced professional art. One may think that when a silk is thrown into the air and is caught by his or her assistant that it has miraculously transformed into a walking cane, but it has not. These are examples of master illusionists, and not of reality.

If God did not develop in this theorized manner, we are left with an alternate development of God by another means which would destroy the necessity of God being the primal actionable force in the cosmos. Therefore, not supreme.

The appearance of God is a much different story, a nascent God was the beginning of creation and that creation was God Himself. How did this happen?

God did say to us, "You will seek Me and find Me when you search for Me with all your heart."[20]

My theory is that over a much longer period than the amoeba happened on earth, God similarly evolved in the cosmos at large. There is a thing in nature called Brownian movement which is the erratic random movement of microscopic particles in a fluid, as a result of continuous bombardment from molecules of the surrounding medium. This principle combined with the science of knowing that there was no particle that did not come from energy means that in the beginning there was nothing but space and energy. However, there is nothing more alive and actively mobile than wavelengths of energy that are always restless when free. The constant movement would, through time, cause enough tangles to begin a formation of mass. Because the energy mass becomes a coalesced energy mass it is an obstacle to free-flowing

---

[20] Jeremiah 29:13

mobile wave lengths. Constant collision with accruement, develops a continually growing mass. The vast array of energy possibilities in the 95.1% that is invisible to us has much more to work with than the 4.9% of energy and matter that God has allowed us to inhabit.

Energy movement obviously is not identical to Brownian movement which happens in our 4.9% corner of the cosmos. Energy movement, however, is restless and in constant motion if not connected to motionless mass. We can't know exactly the mechanism, but with 95.1% of all energy, we can with confidence expect to see various unique surprises from the 95.1% when our mind is able to see the full 100%. It is an easy speculation for energy tangles to stick together in random form until by chance an on/off occurrence happens, making one entity, and allowing for learning bit by bit. Furthermore, nothing prevents the learning curve from becoming exponential.

Eventually the mass would become large enough and alive enough to have spontaneous connections adequate for it to have a nascent awareness of its being constantly bombarded by other bound energy waves or other tangles of wavelengths. That became the first intelligence or consciousness of the entire cosmos. As soon as the consciousness realized that it did not like being hit and first made a thought of volition wishing it to stop: the wish was the word, the word was the volition, the volition caused the action that was the primal action of the Primal God.

Throughout the eons God taught Himself. He learned all of what we see in the universe and all of what is there that he has not yet shown us. Everything that happened in the development, since it is all contained energy, is totally recorded and not one piece of history has been lost, therefore all projects of God are compiled in its library of every thought and action throughout the eons of history, and is available to Him, at His fingertips and His Command.

Skeptics will say energy movement is not the same as Brownian movement and there is an element of truth there. However, we only understand 4.9% of energy, 95.1% can and does have many differences from the 4.9%. God lives and functions in the invisible 95.1% which is indisputable, so this theory has a 95.1% range in which to be viable. We already know that God is developed inside the 95.1% because He is invisible to our five senses and technology.

GOD UNCONSTRAINED

Comments are made that God is immobile, all that can be, the ultimate, therefore unchangeable. I say bunk. To us it might appear so, but even we, lowly in comparison to God, have been blessed with intelligence beyond such ridiculous confines. These thoughts attempt to ascribe constraints to God.

Again, with trillions of light years to develop as I theorize, what possible reason could prevent God from having another idea or action? What would prevent Him from being able to extrapolate His knowledge to another step, just as we humans do regularly? How and why would God develop to this point in time and never grow or develop any further? These are unreasonable confinements of a God we imagine. There are, for example, some religious believers who look forward to getting to heaven, to being given the opportunity to have their galaxy or universe. I am not one.

As soon as God learned that His word, i.e., His will when activated by his volition thought (the word as a command) would put in place the necessary power and control of the surrounding abundant energy to do his bidding and make it happen, He, God, was growing and on his way.

Matured to this point, if He desired, He could transport to wherever He wanted to go. Realistically, a God of whom we speak and describe is likely the functional center of the universe—no matter His location He functions throughout the universe.

These vast distances call into question the ability of God to have instantaneous interaction everywhere. Quantum speed is so fast we cannot measure it. It is instantaneous as far as scientists have been able to measure it. We in our short existence have discovered quantum theory and speed, though we have not yet fully mastered the use of it. We also have learned to clone things, even animals and people, certainly reproduce computers and most everything.

God has such powers, and can at will have as many or as few, if any, substations throughout galaxies and the universe to do his instantaneous bidding. Without question God has that intelligence and ability. We only try to limit His ability to discredit Him, thereby freeing us from responsibility to a higher power than our narcissistic selves.

GOD AS A PERSON

We tend to talk of God as a person. The Bible states that we humans are made in His image. His conversations in the Bible with various people including Adam, Eve and the snake were as a person. Most religions have a name for God as a person.

When we attempt to describe God as a person outside of this world, we seem to know God is not an elder human with a long gray beard. We take for granted that he is a supreme being in a spirit world. How then are we made in His image?

The conundrum is that in order to make sense of this, we must abandon the mindset of anthropomorphic projection. We must look for other evidence of likeness and signs.

God has been compared to a supercomputer. We know however that computers are built out of solid material, cobbled together in a way that it will perform actions directed by the program installed in it. It has fallacies by program error, through damage by the elements, power surge or other abuse. It can do nothing to repair itself. Of course, there are failsafe measures where one item goes down the computer can default to an alternate one, but ultimately it is dependent on others for repair and support. No matter how good the repairs, it will become worn out, obsolescent and useless.

On the other hand, God has all the attributes of a supercomputer but none of the drawbacks. How can this be possible? Because God is a living being. We cannot know the exact formation, but we know He is formed from the essence of life, which is energy, in fact dark energy, and therefore He is alive and living in a sea of energy. He has abundant available energy to use when and however He needs it. We must remain cognizant of—and amazed at—His having to Himself 95.1% of the available energy, not just in our universe, but the entire cosmos and invisible to us, that He may use in many ways to enhance His varied functions of which we cannot even imagine. Furthermore, God's energy is free and is not locked such as the energy in the chair I am sitting in or the car that I drive, or the skyscrapers of Manhattan. They all were, and we are as humans, converted energy. It is locked in a form of matter. Some energy recycles. For example, it is energy that converts to the food we eat, which when digested becomes energy again, then matter, and eventually converted back to energy.

We know that the matter of this 4.9% world deteriorates and does not last. Even metal will have metal fatigue and deterioration. Cells of living organisms don't last the full lifespan of the host. They die as they reproduce. Each time they reproduce the tail of the cell called a telomere is shortened and when it runs to zero the cell cannot further reproduce, and death ensues. Science is working to slow or eliminate the programmed shortening of Telomeres.

The body of God is certainly pure energy. It does not rust, dissolve or otherwise change, except as He directs.

Afterthought:

To illustrate energy being "live," think of the electrical circuits in our house and above us, and below us. If it is powered, we call it a live wire. Energy is the essence of life. Think of what you could do if your entire body was free energy.

What if you had hundreds, thousands, millions, billions, even trillions of years to continue growing, learning, developing, building, etc.

It is plausible that God has chosen to convert His favorite thoughts and images into the human form as His favorite achievement. He honors us saying we are made in His image.

He had no difficulty in allowing His Son to have an image like us, which is quite phenomenal!

Think more analytically about God Himself. His entire being consists of Dark Matter Energy. There is nothing in the dark matter field that is a danger to God, He is in the safest of all places. He is impervious to anything of the Universe. He has given us and at least some animals a Mind made of nothing but dark indestructible matter. Our mind is our soul, the entire repository of our entire being.

He has in fact protected the human race to survive death, arrive in the dark matter part of the cosmos and potentially exist there forever as his guest. How thoughtful, generous and graceful is our God though we can't ignore His rules.

GOD IS THE GOD

Late in the book as often in life, scientists, it seems to me, must admit the only satisfactory answer to the totality of such knowledge becomes so massive, (with numerous observations, detailed recordings of intervening 'cosmic' power including various delivery mechanisms, the totality of which is overwhelmingly, humanly, transparently tangible), that God is the God, yet zero percent understood or explainable with our five senses. However, logic demands we accept divinity—even divine intervention as inescapable. It must be ascribed to a living omnipresent, omni conscious, omni-capable God. No other hypothesis even rises to the level of being a theory.

Believers depend on His grace for our salvation; Christians by way of Christ dying on the cross for salvation for any who believe. If one has never heard of Christ but lives in accordance with the Holy Spirit, the equal third part of the Trinity, that person then is also in God's grace.

RELIGION

Religion is a unique segment of the human experience, just as science or philosophy. It can be, and is, a method for unity, but sadly is sometimes used as a tool of discord and division. Unfortunately, both scenarios thrive. This is not an indictment of any religion, but rather, after a reasonable perusal of comparative religions, an endorsement of religion's intent to find the nature and purpose of our existence. We seek to harmoniously blend our known physical life with a spiritual connectedness, resulting in a spiritual Self, superior to the earthly life. At a minimum we strive to grow in that direction in preparation for the afterlife—the Spiritual existence.

If the foregoing resonates then it would seem the magnitude of the universe, the commonality of humanity, there would be only one creator, but multicultural and ethnic expressions of the human race. We observe this to be the ethnic state of humanity. I theorize this is not to encourage competitiveness but to enjoy diversity.

The foregoing essentially dictates a monotheistic religion inspired by the creator that would be a template for religion. Because of multicultural, ethnic groups, languages, and geographic locations we would expect different interpretations of religion. It is obvious that the founder of each religion would have strong opinions of the importance of their revelation. I and most, I believe, would expect guidance from the Holy Spirit. It is incumbent upon the leaders of each religion to hold the group of believers accountable to the purist spiritual meaning. Some do this better than others.

I believe and propose a theory that God is the parent of religion and was the first prophet of religion.

What do we know and when did we know it? The civilization of Mu existed ten thousand years before Christ was born. Yet, God provided them with the knowledge that something would happen five thousand years after their demise and an additional forty-five hundred years before Christ would be born. In fact, a momentous happening would occur on the day of Christ's actual birth and the day of His crucifixion. Referring to the Hyksos descendants of Mu who built the Great Pyramid and translated this knowledge in scientific language throughout the Great Pyramid, shows us God knew ten thousand years before Christ was born the happenings during those ten thousand years. He knew all of what happened chronologically prior to Christ's birth.

Zarathustra of Iran was the first prophet to move from polytheistic belief to monotheistic belief. His name was pronounced Zoroaster and his religion was Zoroastrian. We know that he lived, and his religion flourished prior to the sixth century B.C. We also know that he prophesied accurately the birth of Christ. In fact, the three wise men were Zoroastrians by faith and following the star prophesied by Zoroaster. Today a few hundred thousand Zoroastrians continue in that faith and belief.

We know from the Bible the Jewish religion believed in a monotheistic God, the coming of Christ, the Messiah. Later controversy developed around the name Messiah.

The life work and miracles of Christ are unparalleled. The crucifixion and resurrection are unequaled. The heavenly signs unparalleled, according to the Bible.

The life and works of Christ are professionally researched and corroborated in secular literature even for the contemporary times of Christ in "Reliable Truth" by Richard E. Simmons III. There are more prophecies regarding God's works and Christ's works than of any other known religion.

There have been benefits and blessings from other religions, reviewing my past from birth to current time. The presence of God has been shown to me through the Christian faith: through doubting times and arduous searching, through taking a college class on the subject, and in my own readings regarding comparative religions. I give reverence and encouragement to any spiritual, uncorrupted divine messages. As noted above, the life and works of Christ are unequaled and unexplainable—except that He is who He claimed to be. Therefore, since He is the number one message delivered from God, the religion He brings and initiates and bears His name, Christianity is the parent religion.

GOD OF HOPE – ACCEPTING HUMILITY

A day is a day. The daily rotation of our earth slows gradually. The last estimate I saw was one millionth of a second per day. This translates to one second every 2739.73 years, omitting calculation of the leap years.

Carbon dating tells us the age of the earth and its developmental stages. Those are facts within narrow margins of error. We have no choice except to believe the days of creation, as written in Genesis, being symbolic language, not accurate per 24-hour day scheduling. Genesis is entirely correct chronologically speaking in the formation.

God, being born without parents, was developed by pure logic. He learned one step at a time through trial or experimentation as to whether something worked or not. That part was strictly electrical as an on-off or yes-no function. Over time knowledge and experiments became much more sophisticated and powerful.

It seems logical that that type of system would have no way to jump from 'yes-no' to complicated things such as feeling, emotion, anger, rage, love. These things must have evolved in the animate creations. The Creator observed and studied them, learned them and incorporated them into Himself, analyzed the good and the bad in those emotions. He famously embraced these emotions by declaring, "I am a jealous God."

God might be looking at humans as becoming hopeless. On the other hand, we cannot know how He sees us, and where He can see that we will eventually land in His sight—victoriously surviving or totally extinguishing ourselves.

His displeasure could be inching toward giving up on us. I choose to not yet be so negative but rather optimistic. There is significant movement in the faith-based people—such that there could be a resurgence of faith and growth in spirituality. The power of prayer is discussed elsewhere.

Since God has total memory and recall, His projections on outcomes would be infallible, except we know He is subject to changing His mind.

There are those who say that God is the ultimate, and is therefore locked into His position, wise and learning. I dispute that. Any thinking that limits God is flawed in that He is all powerful. Those thinkers remind me of the congressman that made a motion to close the patent office when the telegraph was invented. He thought nothing could ever be more innovative in distant communication than the telegraph. Others have even continued to go beyond what we thought were our limits even having flown to the moon and back, and are now contemplating even more distant ventures. If we are not out of ideas why would we believe that God is out of ideas.

God is loving and caring. He responds to prayer, although perhaps not always the way we want. He responds to numbers. The more prayers offered up the heavier the impact on Him. The condition of our country now—and indeed the world—is not good. We do not praise and thank God and Christ adequately. I propose now that every believer and doubter earnestly pray every day during the noon hour, praising God and Christ with thankfulness for our blessings. And for submission to His will to be done, including a spiritual resurgence to continue as a God focused nation of people to bring about a true nation of Spiritual values working toward a heaven on earth. Each person should add their personal touch. Please do not ignore this call to prayer. It could possibly be our last best hope.

It is time for humility. Admit the strength of the forces attacking our faith-based society and nation and the world. We should prayerfully call for God's Divine intervening assistance. Without it, we could possibly be overpowered.

I hope to have shortened your search for truth by taking you along with me through mine. My sincere prayer is that this collection of miraculous stories, scripture and scientific facts together have provided you enough evidence to believe that God is real, is active in our lives, and is available to us now, just as He was in Bible times, and is still in the business of performing miracles, for those who have eyes to see, and ears to hear.

ACKNOWLEDGMENTS

Kathy Logue and extended family
Peggy Holly
My Office Staff
N.S. Xavier, M.D.
Marilyn Schultz
Danna Hawkins
Kay McLean
Sam Saxon, PhD
Wayne W. Krug, PhD
Douglas D, Webster, PhD
Mary J. Henderson

With grateful thanks to those who shared their stories but wished to remain anonymous for confidentiality.

A common bond we share is a passionate belief in and love of God, and a hope that our true stories will assist doubters to find the truth we know and share.

~ H.E. Logue

FURTHER READING

More about Dr. Ed Logue: nsxavier.tripod.com/loguemd

There are many interesting books for further reading. Below are the books and documentary which are referenced in the book.

Documentary:

"The Worlds Most Enlightening Region"

Books:

The Holy Bible

"The Great Pyramid: Man's Monument to Man"
By Tom Valentine
Pinnacle Book, Inc

"The Great Pyramid: Its Divine Message"
By David Davidson
Kissinger Publishing, LLC

"The Two Faces of Religion"
By Dr. Narithookil S. Xavier

"The Far Far Future of Stars"
Donna Goldsmith, Scientific American,
Vol. 23, No. 3

"Miracles: What They Are, Why They Happen, and How They Can Change Your Life."
By Eric Metaxas

Edgar Cayce

Made in United States
Troutdale, OR
09/24/2026

54818127R00196